INDOOR RADON AND ITS HAZARDS

INDOOR RADON AND ITS HAZARDS

Editors

David Bodansky
Maurice A. Robkin
David R. Stadler

Contributing Authors

David Bodansky
Peter A. Breysse
Joseph P. Geraci
Kenneth L. Jackson
Ahmad E. Nevissi
Maurice A. Robkin

University of Washington Press
Seattle and London

Printed in the United States of America

Library of Congress Cataloging-in-Publication Data

Indoor radon and its hazards.

Revision of a report issued by the University of Washington's Environmental Radiation Studies Committee.

Includes bibliographies and index.

1. Radon—Toxicology. 2. Air—Pollution, Indoor—Hygienic aspects. 3. Radon—Environmental aspects. 4. Carcinogenesis. I. Bodansky, David. II. Robkin, M.A. III. Stadler, David R. IV. University of Washington. Environmental Radiation Studies Committee. [DNLM: 1. Environmental Pollutants—adverse effects. 2. Lung Neoplasms—etiology. 3. Radon—adverse effects. WN 300 I41]

RA1247.R33I53 1987 363.7′392 87-16134
ISBN 0-295-96516-9
ISBN 0-295-96517-7 (pbk.)

PREFACE

The goal of this book is to provide an introduction to the indoor radon problem in a form which is suitable both for readers who have no background in radiation studies and also for those who have a familiarity with environmental radiation issues in general but not with details of the radon problem itself. We believe it would also be useful as a supplementary reference in courses concerned with environmental radiation.

The manuscript grew out of a faculty Workshop on Indoor Radon, sponsored by the Environmental Radiation Studies Committee at the University of Washington. Talk manuscripts which had been prepared for the Workshop evolved into a Report issued by the Environmental Radiation Studies Committee under the title of Indoor Radon. This book represents a revision of that Report. While each of the chapters is the responsibility of the individual author or authors, the other participants have contributed suggestions as the manuscript proceeded through successive drafts. Thus each of the authors is indebted to his colleagues.

We are also indebted to many other individuals for suggestions they have made concerning the manuscript or for information which they have provided to us. We wish particularly to thank Bernard Cohen, Fred T. Cross, Henry Hurwitz, Jr., Thomas Gerusky, Robert Mooney, Anthony Nero, James Reiley, Margaret Reilly, Jonathan Samet, and Thomas Sibley, as well as (anonymous) reviewers for the University of Washington Press. We are very grateful to Marion Joyner and Lisa Bickeböller for their skill and dedication in the preparation of the manuscript. The Graduate School Research Fund at the University of Washington has provided funding for the manuscript preparation through its support of the Environmental Radiation Studies Committee.

The Editors

CONTRIBUTORS

David Bodansky, *Professor of Physics, University of Washington*

Peter A. Breysse, *Associate Professor of Environmental Health, University of Washington*

Joseph Geraci, *Associate Professor of Environmental Health, University of Washington*

Kenneth L. Jackson, *Professor of Environmental Health, and Chairman, Radiological Sciences, University of Washington*

Ahmad E. Nevissi, *Research Associate Professor of Fisheries, University of Washington*

Maurice A. Robkin, *Professor of Nuclear Engineering and Environmental Health, and Director, Nuclear Engineering Laboratory, University of Washington*

David R. Stadler, *Professor of Genetics, University of Washington*

CONTENTS

Chapter 1

OVERVIEW OF THE INDOOR RADON PROBLEM

David Bodansky

A. Introduction

The possibility of cancer induction due to indoor radon has been attracting increasing attention in the scientific community during the past decade. It is now widely recognized that indoor radon is the largest single source of exposure to ionizing radiation in the environment. For the population as a whole, the average effective radiation dose from radon is estimated to be greater than the dose from all other natural sources of radiation combined, greater than the dose from medical treatments including x-rays, and very much greater than the dose from industrial activities including nuclear power. As an indication of the possible health impact, the Environmental Protection Agency reports that "scientists estimate that from about 5,000 to about 20,000 lung cancer deaths a year in the United States may be attributed to radon" (EPA 1986a).

Nonetheless, until very recently there has been relatively little awareness of radon outside specialized scientific circles. The purpose of this book is to clarify the nature and significance of the radon issues and the extent to which they constitute a "problem." An overview is presented in this chapter, with details reserved for succeeding chapters.

B. Scientific and public awareness

Ionizing radiations from cosmic rays and from naturally radioactive elements such as radium have always been present on earth, but our awareness of their existence dates, at most, to about the turn of the century, when both x-rays and natural radioactivity were discovered. By the mid-1920s, there was an active realization that such radiations could be harmful, at least at the high exposure levels reachable in the medical use of x-rays and through contact with strong sources of natural radioactivity, especially radium extracted from uranium ores. During and after World War II, nuclear fission and its radioactive products became a cause for still greater concern.

From the 1920s and intensifying after World War II, substantial attention was devoted by the scientific community to understanding the effects of these radiations and to controlling the levels of exposure from the recognized sources. Awareness of the magnitude of the exposures from indoor radon, and consequently interest in radon, lagged far behind. Although it had been recognized since the 1950s that high radon levels in uranium mines caused an elevated incidence of lung cancer among miners, only since the late 1970s has the issue of radon exposures of the general public been extensively addressed.

This situation has now changed. The National Council on Radiation Protection and Measurements (NCRP), the national body charged with providing scientific advice on radiation matters, made radon the topic of two major reports published in 1984 (NCRP 1984a, 1984b). *Health Physics*, the official publication of the Health Physics Society, devoted its August 1983 issue to the topic of "Indoor Radon" and that topic is now one of the leading preoccupations of the journal. Numerous scientific meetings during the past few years have either focused entirely on indoor radon or have made radon a major subtopic within the broader context of indoor air quality.

This scientific interest has been paralleled by increased activity on the part of governmental agencies. The Environmental Protection Agency and the Department of Energy have been somewhat concerned about the indoor radon problem since the late 1970s, and the EPA now views it as "definitely the biggest environmental radiation problem in the United States today" (Guimond 1986). To provide the public with more information on radon, the EPA issued in 1986 two widely publicized pamphlets: *A Citizen's Guide to Radon: What It Is and What to Do About It* (EPA 1986a) and *Radon Reduction Methods: A Homeowner's Guide* (EPA 1986b).

Some indication of the possible magnitude of the radon problem had been given in a 1985 report of an inter-agency federal group, which included Environmental Protection Agency and Department of Energy representatives, which stated (CIAQ 1985):

> If these risk estimates [of lung cancer induction from exposure to radon] are borne out by further research, several thousands of deaths per year in the U.S. population can be attributed to indoor radon. The estimated upper limit of health effects is 30,000 deaths per year.

The later EPA estimate of about 5,000 to about 20,000 annual deaths (EPA 1986a), cited above, essentially repeats these figures, although with a slightly more confined quoted range.

Despite this concern, funding for research and remedial action on radon is at a relatively low level, especially when contrasted with the efforts which are being devoted to other radiation-related safety issues. The EPA supports a modest program to study radon problems, but most of the cost and responsibility for radon research, monitoring and mitigation are being left to the states (NYT 1986). As of early 1987, there were no federal regulations to control radon exposure in the home and there are no federal funds for assisting individual homeowners in reducing radon levels, aside from research projects. As discussed below, the EPA has put forth suggested standards for indoor radon (EPA 1986a), but these are not mandatory standards and have no legal force unless individual states choose to embody them in new regulations. The states vary widely in their level of concern about radon, in part because there are strong regional differences in radon levels.

Public interest in radon has been occasionally piqued by articles in the general press. Considerable attention has been given to the high radon levels that were uncovered in the Reading Prong region of Pennsylvania, following the discovery in late 1984 of extremely high levels in one home. This discovery was precipitated by radiation alarms that were set off by an engineer upon entering the Limerick nuclear power plant. The radioactive contamination which triggered the alarm was traced to natural radon in the engineer's home, not to anything connected with the nuclear power plant. Radon exposures received by people in that house were found to be about 100 times greater than exposures permitted for workers in uranium mines. Other homes in this region were subsequently found also to have elevated radon levels, although usually far lower than those in this extreme case.

However, despite so striking an event, the issue has remained fairly quiet in terms of broad public concern. A succinct explanation for the relative lack of attention has been offered in the *London Observer* (Lean 1986):

> Radon exposure is the hottest issue among radiation scientists today—but it remains almost entirely unknown to the public. Neither environmentalists nor the nuclear industry seem to like talking about it.
>
> Environmentalists keep quiet because concern over radon in houses would divert attention from their campaign against nuclear power. The nuclear industry, for its part, does not seem to enjoy publicizing the ill-effects of radiation in any form.

This explanation may be close to the mark, although of course there are exceptions, both among environmental groups (see, e.g., EDF 1986) and among supporters of nuclear power (see, e.g. Hurwitz 1983 and Cohen 1980).

The true seriousness of the indoor radon problem is not known at this time. While the impact could conceivably be as high as the upper EPA value of 20,000 annual deaths (even higher if one accepts some figures presented in EPA 1986a), it also could be substantially below the lower estimate of 5000 annual deaths. The latter would be the case if, as believed by some, ionization radiation is less harmful at low and moderate doses than is assumed in standard calculations. However, whatever the hazards of radiation, it is well established that radon is a very major radiation source. As such, the nature of the radon problem deserves to be better and more widely understood than has hitherto been the case.

This is not just a matter of so-called academic interest. Unlike the situation for exposures from the radioactive isotopes found naturally in the body, such as potassium-40, it is possible to take actions which may increase or decrease the radon exposure levels. Indoor radon concentrations can be increased by lowering air infiltration rates, in the absence of compensating steps. They can be decreased by reducing radon entry into houses or by favorable ventilation patterns. The nature of the energy conservation measures taken and the way in which houses are built in the future will be influenced by our understanding of the relation between these measures and the broad problems of indoor air pollution, including specifically the radon problem.

C. What is radon?

Radon is a radioactive gas, first discovered in the early 1900s. During the early studies of radioactive elements at the turn of the century, it was found that gaseous "emanations," themselves radioactive, were associated with many of the newly identified radioactive elements. The gas associated with uranium and radium was called radon. Subsequently, in precise usage, the term radon has come to designate the element with atomic number 86. The original "radon" is now known to be radon-222, the most abundant of the several isotopes of the element radon. However, in common usage, including discussions of the "radon problem," the term is still used to denote radon-222 itself.

Chemically, radon is a noble gas. As such, it is similar, for example, to helium and neon. These gases do not readily interact chemically with other elements and are relatively difficult, although not impossible, to trap. Like any other noble gas, radon is colorless and odorless. If it is in the air, it is inhaled along with all other gases. It is also exhaled promptly, and were one dealing with radon alone there would be little reason for concern.

The radon hazards do not come primarily from radon itself, but rather from radioactive products formed in the decay of radon-222. These products, called the "radon daughters," are also radioactive but, unlike radon, they are atoms of heavy metals and readily attach themselves to whatever they contact. The main health problems stem from the inhaling of radon daughters, or dust particles carrying radon daughters, and the subsequent lodging of the radon daughters in the lung. Nonetheless, the problem can be properly called the "radon problem," because the daughters are present only in environments into which the radon has penetrated.

D. Origin of radon in the environment

Radon-222 is the direct product of the decay of the most prominent radium isotope, radium-226, which in turn is a product, several steps removed, of the decay of the most abundant uranium isotope, uranium-238. The uranium is believed to have originally been produced in the supernovae explosions responsible for the nuclear synthesis of all of the heaviest elements found in nature. The products of these and other cosmic synthesis processes were combined in the cloud out of which the earth condensed some 4.6 billion years ago. Those radioactive nuclei in the cloud with half-lives of less than a few hundred million years have virtually all disappeared by now. This leaves the stable nuclei, which make up most of ordinary matter, as well as the very long-lived radioactive nuclei, such as potassium-40 and uranium-238, together with their decay products.

The half-life of uranium-238 is 4.5 billion years and today, therefore, about one-half remains of the uranium-238 present when the earth was formed. The decay of uranium-238 is the first step in a chain of successive decays of radioactive isotopes, with radium-226 and radon-222 as intermediate products. The decay chain finally terminates when a stable isotope, lead-206, is reached.

This decay chain is shown in detail in Table 2-1 of Chapter 2. From the standpoint of the radon problem, the decays of importance are those in the middle range, from radium-226 to lead-210. Omitting minor sidepaths (involving less than 0.1% of the decays) these decays are:

Ra-226 →	Rn-222 →	Po-218 →	Pb-214 →	Bi-214 →	Po-214 →	Pb-210
1600 y	3.82 d	3.11 m	26.8 m	19.8 m	0.0002 s	

The first line above specifies the nuclei in the chain, and the second their half-lives, in years (y), days (d), minutes (m), or seconds (s). The standard chemical symbols are used: Ra for radium, Rn for radon, Po for polonium, Bi for bismuth, and Pb for lead. The term "radon daughters" is usually used to designate the four nuclei immediately following radon-222. (The fifth nucleus after radon-222, lead-210, is in a different category because it has the relatively long half-life of 22 years.)

Uranium is ubiquitous in the earth's crust, being present in virtually all forms of rock and soil. Thus, its daughter, radium-226, is also present throughout the earth's crust. When an atom of radium-226 decays, an atom of radon-222 is formed. The radon can then diffuse out of the upper layers of the ground into the atmosphere.

Radon-222 has a half-life of 3.82 days, and if radon is held up in the ground for several days or more the diffusion of radon out of the ground is substantially reduced. The fraction of the radon which escapes depends upon the depth at which it is formed and the permeability of the ground to radon. Typically, only radon which originates near the surface will escape. As a rough average guide, approximately 10% of the radon formed in the top meter of soil escapes (NCRP 1984b). The radon daughters attach themselves to matter much more readily than does radon and if a radon atom decays in the ground its daughters will be trapped. Conversely, the radon which escapes into the air spreads its decay products into the general environment.

Typical uranium concentrations in soil and rocks are several parts per million (by weight), with substantial variations in concentration from place to place. When the concentration is as high as several thousand parts per million, the uranium deposit may be attractive for uranium mining. There are many regions where the concentration is intermediate—well above the average levels but still not suitable for mining. If uranium-rich material lies close to the surface of the earth there can be high radon emanation rates. In such locations, the hazards from radon exposures can be much greater than average.

E. Why consider only Radon-222?

The "radon problem" refers almost always just to radon-222 and its daughters. However, other radon isotopes are formed in the decay of other parent nuclei. In particular, the decay of uranium-235 leads to the formation of radon-219 (also known as actinon) and the decay of thorium-232 leads to the formation of radon-220 (also known as thoron). Radon-219 may be ignored entirely, partly because its forerunner, uranium-235, is a relatively rare isotope and partly because the half-life of radon-219 is only 4 seconds, and therefore almost all the radon-219 will decay before it can escape from the ground.

The abundance of thorium-232 in the earth's crust is somewhat higher than that of uranium-238, but because of the longer half-life of thorium-232 the average rate of production of radon-220 in the ground is about the same as that of radon-222. The half-life of radon-220 is 55 seconds, much less than that of radon-222, and there is a greater chance for radon-220 to decay while still trapped in the ground or in building materials than for radon-222. In consequence, the amount of radon-220 entering the environment is less than the amount of radon-222, and radon-220 and its daughters are usually neglected in discussions of the "radon problem" although their contribution is not as trivial as that from radon-219.

F. Indoor radon levels

Radon enters houses primarily from the ground, mostly through gaps in the building structure. It also enters, usually in lesser amounts, from the building materials of the house, from the water supply, and, if used, from natural gas. Normally, indoor radon levels are considerably higher than outdoor levels. If the radon input rate is constant, the radon concentration is determined largely by the ventilation rate. Other things being equal, if the rate of air exchange with the outdoors is halved the radon concentration would be doubled. For this reason, it has been feared that weatherization of houses for energy conservation, including weather stripping and caulking of cracks, would lead to high radon levels.

Subsequent investigations have not shown large effects from weatherization, apparently because the steps taken usually do not greatly reduce the air infiltration rates and in some cases because the radon input rate may be reduced due to increases produced in the basement air pressure. However, despite the importance of determining the impact of weatherization, the actual consequences are not yet well known. Studies to date do not suffice to provide a reliable prediction of the effect of weatherization on the radon level for a specific individual house, nor is there a good quantitative summary of the spectrum of consequences for the housing stock as a whole.

Nevertheless, it is well established that the differences found in radon concentrations among different houses stem more from differences in the rates of radon entry than from differences in ventilation patterns. Houses differ greatly in the rate of radon production in the underlying soil and in the degree to which the radon can penetrate into the house. In consequence, radon concentrations vary widely. The approximate average (arithmetric mean) concentration for single-family houses is 1.5 picocuries of radon per liter of air (pCi/ℓ). In an extensive survey of representative data (Nero 1986), most of the single-family residences were found to have radon levels lying within a band extending from about 0.2 pCi/ℓ to 4 pCi/ℓ, with about 9% having higher concentrations. About 2%, corresponding to about one million homes, have levels in excess of 8 pCi/ℓ. This concentration is at twice the EPA's recommended threshold for remedial action, namely 4 pCi/ℓ (EPA 1986a).

These variations by no means encompass the full range of variation. In the Reading Prong region of Pennsylvania, the region of particularly high concentrations cited above, 12% of the houses had concentrations above 20 pCi/ℓ (Gerusky 1987), corresponding to exposure levels higher than those considered acceptable for uranium miners. This does not appear to be an unique situation (Cohen 1986), but much remains to be done to characterize the distribution of radon levels over the country as a whole.

Efforts to control radon concentrations in houses have been made in a number of ways, including: (a) elimination of cracks and other entry channels in the basement, (b) ventilation of crawl spaces or basements, (c) removal of radon daughters by filtration or electrostatic precipitation, (d) ventilation of the house with use of air-to-air heat exchangers or by opening windows, and (e) drawing radon away from the house externally either at drain tiles or, when the house is built on one, from underneath the concrete slab. The effectiveness of these methods remains a matter for continued investigation, but it is clear that decreasing the rate of entry from the ground is of major importance. These and other techniques are discussed briefly in the guide prepared by the EPA (1986b) and more extensively in the EPA report (1986c): *Radon Reduction Techniques for Detached Houses, Technical Guidance.*

G. Procedures in estimating the health impact of radon

Inhalation of radon, or more specifically the radon daughters, leads to deposition of radioactive atoms on the walls of the lung, especially in the bronchial region. In the decay of these atoms, alpha particles are emitted which irradiate the cells of the lung tissue through which they pass. These irradiated cells may become cancerous.

The rate of lung cancer induction depends upon the kind, number, and location of the radioactive atoms deposited in the lung. These in turn depend upon many factors, including the radon concentration in the air, the ratio of the concentration of the daughters to that of radon itself, the extent to which the daughters are attached to dust particles and the sizes of the dust particles, the breathing rate, and the rate of deposition of the daughter atoms in various parts of the lung. Given data for these factors and an adequate model of the lung, it is possible to calculate the radiation dose. A number of extensive calculations of this sort have been carried out, and while their results are not in complete agreement, they are close enough to permit reasonably good estimates of the radiation dose to the lung resulting from a given radon concentration in the air.

The dose itself provides one indication of the health impact of radon, especially when comparisons are made to doses from other radiation sources. A more direct, although more uncertain, measure is the number of lung cancers attributable to radon. (Lung cancer is usually fatal, so there is little distinction between lung cancer incidence and lung cancer deaths.) At present, estimates of the relationship between lung cancer and the magnitude of the radon exposure rely primarily on extrapolations from the past experience of miners, in uranium and other mines. The high lung cancer rates for these miners were found to be correlated with high exposures to radon in the mines. The observed correlations for miners are translated into a calculated rate of lung cancer increase per unit of radon exposure. It is then assumed that this rate (i.e., cancers per unit dose) applies not only at the high radon exposures of uranium miners but also at the low radon exposures typical of most homes.

There are many uncertainties in this procedure and in the mortality estimates which result from it. In the first instance, it is difficult to establish unambiguously the radon exposures suffered by miners over their working lifetimes as well as the possible contribution to lung cancer of other carcinogens to which the miners may have been exposed. Thus, the relationship between lung cancer incidence and radon exposure is somewhat uncertain even for miners. A second uncertainty in the calculation stems from differences in the conditions experienced by miners and by the general population (e.g., differences in the dust in the air and in breathing patterns), so that identical radon concentrations in mines and in homes may produce different radiation doses to the lungs. In addition, there may be a synergism between radon and smoking but the extent of the synergism, if any, is not well known.

Beyond these uncertainties, and probably overshadowing them, is the uncertainty in the premise which underlies the extrapolation from the doses received by miners to those experienced in homes, namely that the number of lung cancers is directly proportional to the radiation dose. This assumption is known as the "linearity hypothesis" (see Chapter 8, Section C). The validity of the linearity hypothesis is a matter of continuing debate. There is no substantial empirical evidence establishing its validity for humans at low radiation doses, and there is some evidence from studies of cancer incidence among miners that the cancer incidence rate per unit exposure is less at low dose rates. Further, in a number of recent studies of lung cancer incidence in regions with elevated indoor radon levels, no evidence was found of the increase which would be expected were the linearity hypothesis valid for radon exposures (Cohen 1987, Hofmann 1986).

These studies are not generally taken to be conclusive, and uncertainty as to the applicability of the linearity hypothesis remains. A sizable number of scientists believe that the application of the linearity hypothesis at low dose levels leads to an overestimate of the true effects; others believe that it underestimates the true effects. Overall, at least for alpha particles, the hypothesis serves as the present consensus estimate, and radiation regulatory bodies throughout the world rely on the linearity hypothesis as a prudent basis for establishing protective standards. For the future, there is hope that studies of correlations between radon levels and lung cancer incidence in the general population will help shed light on the validity of the linearity hypothesis (Cohen 1987).

H. Magnitude of the health impact of radon

How bad is the radon problem, granted the uncertainties? The answer can be given in two parts. First, if we compare radiation exposure from different sources on the basis of an equivalent whole-body dose, indoor radon is estimated to be the greatest single source of ionizing radiation exposure experienced by the U.S. population. Second, radon is responsible for only a small fraction of lung cancer incidence. These statements are not in contradiction because, with existing exposure levels, ionizing radiation is a relatively minor cause of cancer.

It is not yet possible to put these summary statements in precise quantitative terms, but it is possible to obtain a helpful quantitative perspective. Current surveys suggest that average radon concentrations are in the neighborhood of 1.5 pCi/ℓ for single-family houses. (In a different, commonly used, terminology [see Chapter 2] a radon concentration of 1 pCi/ℓ is usually taken to correspond to a radon daughter concentration of about 0.005 working level [WL], and to a total exposure to the lung of about 0.2 working level months [WLM] per year.) The resulting radiation dose is roughly equivalent in terms of overall detrimental effects, especially cancer deaths, to a whole-body radiation dose of about 300 millirem (mrem) per year. This is a reasonable "best" estimate of the average effective radiation dose. With a different choice of numerical values it is possible to get plausible alternative estimates for the average radon dose in the United States, ranging from 100 mrem per year to 800 mrem per year. For individual homes, the annual doses vary much more, from well under 100 mrem to many thousands of millirem.

The effective whole-body radiation dose from radon can be directly compared to whole-body doses from other sources. The average total annual dose from natural sources, including cosmic rays, radionuclides in the ground, and natural radionuclides in the body, is about 100 mrem. The average annual medical exposure, where about half the population receives effective doses approaching 200 mrem, is also commonly taken to be about 100 mrem. Thus, not only is radon the largest single source of radiation exposure, averaged over the entire population, but for much of the population its contribution exceeds the total from all other sources.

As discussed above, there are large uncertainties in the estimates of the number of lung cancers produced by radon exposures. Some of these uncertainties are embraced in the EPA estimate that radon may currently be causing between 5,000 and 20,000 deaths per year in the United States. There seems to be little possibility of obtaining a more precise estimate at the present time and, if the full range of uncertainties is included, the estimate could run from near zero to about 50,000. The figure of 50,000 deaths per year corresponds to extreme risk numbers put forth by the EPA along with the more conservative 5,000 to 20,000 range cited above (EPA 1986a). However, at least if one considers the effects of radon alone, without smoking, it appears unlikely that the true number is appreciable more than 20,000 per year (see Chapter 9), and this may be taken as a rough upper bound. The lower number cited in the EPA range, 5,000, is equally plausible. In fact, radon may cause many fewer than 5,000 deaths per year, because even the estimate of 5000 is based on the linearity assumption which may lead to a substantial overestimate of the hazards.

It is considered prudent for purposes of radiation protection to assume that linearity holds, especially for heavily ionizing particles such as alpha particles. If this assumption is accepted, a reasonable estimate is that the United States cancer toll from radon is about 10,000 deaths per year, within a factor of two, and this number is commonly quoted in discussions of the radon problem. Although this is a large number, it still represents less than 10% of total lung cancer deaths and only about 2% of all cancer deaths. Thus, if one focuses on radiation per se, radon is a major problem. If one focuses on cancer itself, radon is a secondary one.

Radon is not evenly distributed among homes, as has been discussed in Section F. For the one million or so homes with levels above 8 pCi/ℓ (about 0.04 WL), the annual exposures correspond to a whole-body equivalent of roughly 1500 mrem. Still greater exposures are encountered in individual cases. In an extreme case, a radon daughter level above 10 WL was found in the initial Reading Prong home (Gerusky 1987), corresponding to about 400,000 mrem per year for a person spending most of the time in this house. An annual exposure of this magnitude is thousands of times greater than the highest total exposure received by any individual (excluding workers) from the Three Mile Island nuclear accident, also in Pennsylvania.

Even at the same level of exposure to radon daughters, radon may not affect all people equally. In particular, although the issue is not settled, there is evidence to suggest that there may be a synergism between radon and smoking (see Chapters 8 and 9). Thus, the impact of radon and smoking together may be greater than the sum of their individual impacts. A given increase in the radon concentration would then lead to a greater increase in the lung cancer rate among smokers than among non-smokers, and concern over radon would become intertwined with the broader concern over smoking.

I. The regulatory dilemma

The guiding principles for the regulation of radiation exposure are that there is no level of exposure below which one can be certain that there will be no adverse health effects when a large population is exposed, and that therefore exposures are to be kept to levels which are as low as reasonably achievable (ALARA). In this spirit, and because the goal is believed to be easily achieved with existing technology, the maximum allowed exposure for any member of the public from nuclear reactors or future nuclear waste disposal sites is limited to 25 mrem per year. For other federally operated or licensed facilities, the limits range from 25 to 100 mrem per year. In fact, these levels are rarely reached. Even the accident at Three Mile Island, while exceeding the regulatory limit for nuclear reactors, did not lead to any individual off-site exposures as high as 100 mrem.

Radon exposures are quite a different matter. Requiring that individuals modify their houses to reduce radon levels would at present be viewed as an unacceptable infringement of personal freedom, and in any event it would be prohibitively expensive to bring all homes down to whole-body equivalent radon exposure levels of, say, 100 or 200 mrem per year. In a few instances, help has been offered to modify houses with unusually high levels, as in the Reading Prong region, but overall little has been done beyond the setting forth of recommendations. Thus the National Council for Radiation Protection and Measurements recommends that remedial action be taken if the annual exposure exceeds 2 WLM (corresponding to roughly 2000 mrem per year) and the Environmental Protection Agency recommends that remedial action eventually be taken if the radon concentration is above 4 pCi/ℓ or 0.02 WL (corresponding to roughly 800 mrem per year). But these are only recommendations. Few individuals know that there may be a radon problem and fewer know the radon level in their homes or what to do about it if it is high.

The regulatory requirements can be further contrasted by considering the cases of radon and nuclear waste disposal. For radon, although implementation of the EPA's suggested remedial actions would reduce the risk for those individuals receiving the highest exposures, it would not even halve the estimated overall toll of 5,000 to 20,000 cancer deaths per year. For a nuclear waste repository, the Nuclear Regulatory Commission and the EPA have established regulatory requirements under which, according to EPA estimates, a repository is "projected to cause no more than 1000 premature cancer deaths in 10,000 years . . . an average of 0.1 fatality per year" (EPA 1985:38069). The contrasting standards (thousands of deaths per year vs. one-tenth death per year) may appear bizarrely inconsistent, but they follow from plausible estimates of what is "reasonably achievable."

The policy issues become even more complex when one considers the question of the weatherization of houses to conserve energy. Because typical weatherization measures to reduce air infiltration are not very effective, the absolute increase in radon concentration is thought to be low, in the neighborhood of only 10 to 20% (Nero 1986). However, there has been little attempt to examine weatherization with the same care and the same criteria that have been applied to other steps which may increase radiation exposure. To what extent should public agencies or other groups which encourage weatherization warn individuals about possible increased radiation risks, or other risks of indoor air pollution, and explain steps which may be taken to reduce these risks?

J. Policy implications

These are difficult issues, all the more so because the dangers of low-level radiation are speculative. The governmental regulatory apparatus is unable to address the issues effectively. Nonetheless, sound public policy suggests certain steps for radon:

1. Indoor radon levels should be mapped through an extensive nationwide program of measurements in homes and working places. This will provide a better characterization of radon levels and hence of the magnitude of the overall problem; it will help identify locations where prompt remedial action is called for; and it will provide data for epidemiological studies of the relation between cancer incidence and radon levels.

2. There should be more extensive studies of means of reducing radon levels for new construction and, where levels are excessive, through the modification of existing homes. Optimal techniques for doing this are not as yet well established.

3. Further studies should be carried out on the relationship between the tightening of houses, as motivated by conservation considerations, and changes in radon levels. If appropriate mitigation steps are taken, there appears to be no conflict between the goals of effi-

cient energy use in homes and low levels of indoor pollution. However, many programs to reduce air losses in houses have shown inadequate attention to potential difficulties.

4. Epidemiological studies should be intensified to establish the relationship between radon and cancer incidence. Radon is unique as a source of elevated radiation doses for large numbers of people. Despite the confounding difficulties, this offers an unusually favorable opportunity for the study of correlations between radiation level and cancer incidence at moderate and low radiation levels. This is important not only for radon but, potentially, for all other aspects of radiation protection.

5. Society's response to radiation exposures from radon and from other sources should be made consistent. We should either be less concerned about exposures from medical treatments and nuclear power or we should be more concerned about the greater exposures from radon. A consistent response would better enable us to decide the acceptability of various radiation risks and to allocate resources for the study and reduction of these risks. A better understanding of the validity of the linearity assumption, through its study in the case of radon exposures, could help in the making of reasoned assessments.

Finally, as a matter of perspective, it is well to note that in terms of the fraction of damage produced radon exposures are not a very large source of environmental hazard, even for lung cancer. A 10% reduction in smoking would probably reduce lung cancer incidence by as large an amount as the elimination of all radon. Radon may be our most important radiation problem, but it is by no means our greatest public health problem.

K. Contents of the succeeding chapters

The issues introduced above will be discussed further in the remaining chapters. In Chapter 2, some of the basic terminology in radioactivity and radon dosimetry is introduced. Methods for detecting radon are discussed in Chapter 3. The levels of radon in the outdoor environment are discussed in Chapter 4. Chapters 5 and 6 are devoted to indoor radon levels, in terms of both what is found and what can be done to change them. The magnitude of the radiation dose to the lung for a given radon level in the environment is discussed in Chapter 7 and the health effects of these doses are discussed in Chapter 8 and 9. Finally, in Chapter 10 comparisons are made between the hazards from radon and from other radiation sources.

References: Chapter 1

CIAQ (Inter-agency Committee on Indoor Air Quality), 1985, "Report of the CIAQ Radon Workgroup," R. J. Guimond and W. Lowder, co-chairs, Radon Working Group, May (Washington,

DC: CIAQ).
Cohen B. L., 1980, "Health effects of radon from insulation of buildings," *Health Phys.* **39**, 937.
______, 1986, personal communication.
______, 1987, "Tests of the linear-no threshold dose-response relationship for high-LET radiation," *Health Phys.* **52**, 629.
EDF (Environmental Defense Fund), 1986, "Growing concern about radon exposure in U.S.," *EDF Letter*, May, p. 7.
EPA (Environmental Protection Agency), 1985, "Environmental Standards for the Management and Disposal of Spent Nuclear Fuel, High-Level and Transuranic Wastes; Final Rule," 40 CFR Part 191, *Federal Register* **50**, 38066 (September 19).
______, 1986a, published with the Department of Health and Human Services, *A Citizen's Guide to Radon, What It Is and What To Do About It,* OPA-86-004 (August).
______, 1986b, *Radon Reduction Methods: A Homeowner's Guide,* OPA-86-005 (August).
______, 1986c, *Radon Reduction Techniques for Detached Houses, Technical Guidance*, EPA/625/5-86/019 (June).
Gerusky T. M., 1987, "Pennsylvania: Protecting the homefront," *Environment* **29**, No. 1, 12.
Guimond R. J., 1986, quoted in *The New York Times,* August 15, p. 8.
Hofmann W., Katz R. and Zhang C., 1986, "Lung cancer risk at low doses of alpha particles," *Health Phys.* **51**, 457.
Hurwitz H., Jr., 1983, "The indoor radiological problem in perspective," *Risk Analysis* **3**, 63.
Lean G., 1986 "Radon: second largest cause of lung cancer," *The Observer* (London), July 13, p. 49.
NCRP (National Council on Radiation Protection and Measurements), 1984a, "Exposures from the Uranium Series with Emphasis on Radon and Its Daughters," *NCRP Report No. 77* (Bethesda, MD: NCRP).
______, 1984b, "Evaluation of Occupational and Environmental Exposures to Radon and Radon Daughters in the United States," *NCRP Report No. 78* (Bethesda, MD: NCRP).
Nero A. V., Jr., 1983, "Indoor radiation exposures from ^{222}Rn and its daughters: a view of the issue," *Health Phys.* **45**, 277.
Nero A. V., Schwehr M. B., Nazaroff W. W. and Revzan K. L., 1986, "Distribution of airborne radon-222 concentrations in U.S. homes," *Science* **234**, 992.
NYT (New York Times), 1986, "EPA Says Radon Is State's Battle," August 16, p. 8.

Chapter 2

TERMINOLOGY FOR DESCRIBING RADON CONCENTRATIONS AND EXPOSURES

Maurice A. Robkin

A. Introduction to radioactivity and radiation

Each atom of matter is composed of a nucleus surrounded by a cloud of electrons. An atom is called radioactive if its nucleus can emit one or more energetic radiations.

Starting with a collection of identical radioactive atoms, the time it takes for one-half of them to decay is called the half-life. After two half-lives, only one-fourth of the original collection is left (i.e., $1/2 \times 1/2$). The decay product, or daughter atom, may also be radioactive, in which case it will decay into another radioactive daughter. This process will continue until finally a stable, i.e. non-radioactive, daughter is reached. Each radioactive daughter has its own characteristic half-life and radiations.

The decay rate has traditionally been specified in curies (Ci). The original curie unit was based on the decay rate of one gram of radium-226, which is approximately 37 billion disintegrations per second (dps). The curie is now defined to be exactly this rate. It is often convenient, particularly in discussing radon, to introduce a smaller unit, the picocurie (pCi), where 1 pCi $= 10^{-12}$ Ci . In international usage, a new unit, the Becquerel (Bq), has recently been adopted. This unit is part of the Système International de Unités (S.I.) and is equal to one disintegration per second (dps) or about 27 pCi.

The ability of a particle emitted in radioactive decay to cause damage is related in large part to its kinetic energy. The kinetic energy is usually expressed in units of electron-volts (eV). An electron volt is the amount of kinetic energy an electron would receive if it were accelerated through a potential difference of one volt. It is a very small amount of energy. If each electron had a kinetic energy of one eV, it would require 2.6×10^{19} electrons depositing their energy in one gram of water to raise the temperature of the water by one Centigrade degree. In nuclear physics, energies are often expressed in million-electron-volts (MeV), where 1 MeV $= 10^6$ eV.

B. Types of radiation emitted from radioactive atoms

Three different radiations were identified in the early studies of radioactivity. These were called alpha rays, beta rays and gamma rays. After the properties of these radiations were better established, they were referred to as alpha particles, beta particles, and gamma rays. In terms of the radiation hazards from radioactive materials in the environment, these are the only radiations of interest. Their properties are briefly described below.

Beta Particles: One of the commonest transformations of a radioisotope is by the emission of a light, negatively charged particle from its nucleus. This particle, now known to be an electron, is called a beta particle. Beta particles are emitted from the nuclei of both light and heavy radioactive atoms. For example, both tritium (hydrogen-3) and lead-214 emit beta particles.

Alpha Particles: Some very heavy atoms transform by the atom emitting from its nucleus an energetic, positively charged particle, many thousands of times heavier than an electron. This relatively massive particle is called an alpha particle. It is now known to be the nucleus of a helium atom. When the alpha particle loses its kinetic energy, it is neutralized by attracting electrons and becomes a normal atom of the element helium. Uranium-238, radium-226, and radon-222 are typical alpha-particle emitters.

Gamma Rays: In many cases, the emission of alpha or beta particles leaves the new nucleus in an excited energy state. The excited nucleus loses energy by prompt gamma-ray emission. The properties of gamma rays are similar to those of visible light, except that the gamma-ray photons have energies that are typically about one million times greater than the energy of the photons of visible light. Typical radioisotopes which emit gamma rays following beta particle emission are cesium-137 and iodine-131. (Some radioisotopes, such as strontium-90 and tritium, emit beta particles but not gamma rays. However, this sort of beta decay is not the usual case.)

C. Radiation penetration

An alpha or a beta particle of a particular energy has a definite distance, called the "range", of penetration in materials. A typical beta particle has a range of about 0.15 centimeters (cm) in soft tissue, which is sufficient for it to penetrate the epidermis and reach the living dermal tissue beneath. A typical alpha particle can only penetrate about 0.005 cm of tissue and so cannot go through the epidermis. Isotopes which emit alpha particles are a hazard only when they are ingested, inhaled or introduced into the body through a wound. Beta-particle emitters are hazardous if they are introduced into the body, if they are spread on the skin, or if they are close by and unshielded.

Gamma rays can penetrate deeply into materials so that gamma-ray emitters are hazardous for either external or internal irradiation. The degree of hazard for a given radiation depends on the amount of radioactivity, the energy of the radiation, the location or part of the body being irradiated, and the type of radiation.

D. Units of radiation dose

Radiation can interact with matter and the damage caused by the radiation is correlated with the amount of energy deposited per gram of material through which the radiation passes. The energy is expended in removing orbital electrons from around the atoms of the material, a process called "ionization". This energy deposition is called the "physical dose" or simply the "dose". The unit used for the dose is the "rad" in the traditional centimeter-gram-second system of units and the gray (Gy) in the S.I. system of units. One Gy is defined to be the deposition of 1 joule of energy in 1 kilogram of material while 1 rad is defined to be the deposition of 100 ergs of energy in 1 gram of material. One Gy equals 100 rads. It would require over 4000 Gy to raise the temperature of soft tissue by 1 Centigrade degree. Biological damage due to irradiation is the result of ionization rather than temperature rise.

In the case of biological tissue, the damage is correlated both with the physical dose to the tissue and with the density of ionization that the radiation causes in the tissue. The density of ionization determines how concentrated is the energy deposition (on a microscopic scale). Different radiations produce different densities of ionization for the same physical dose. In particular, alpha particles cause much higher ionization densities than do beta particles or the energetic electrons generated in material by gamma ray interactions.

The combined effect of physical dose and ionization density is called the "dose equivalent". The unit used for the "dose equivalent" is called the "rem" in the c.g.s. system and the sievert (Sv) in the S.I. system. 1 Sv equals 100 rem. Equal numbers of rems (or sieverts) are intended to imply equal amounts of biological hazards. The dose equivalent is obtained from the physical dose by multiplying the latter by a dimensionless scale factor equal to the product of various dimensionless factors. Each factor represents a modification of the effectiveness of the particular radiation for causing biological damage due to variations in the particular physical situation. For example, a modifying factor may be required because of high density of ionization or of non-uniformity in the distribution of radioactivity in the body. The most common modifying factor, called the "quality factor", Q, corrects for the differences due to the density of the ionization caused by the particular radiation. For example, for the energies of alpha particles emitted by radon daughters, the quality factor is often taken to be 20 while for gamma rays and beta particles it is unity. Thus, one rad of dose from alpha particles would be equal to 20 rem and would represent 20 times the biological effect of one rad of gamma rays (equal to 1 rem). Similarly for $Q = 20$, it would take 20 rads of gamma radiation (20 rem) to give the same biological effect as 1 rad of alpha radiation (20 rem).

E. Radon and its daughters: secular equilibrium

Radon-222 is a colorless, odorless, noble gas which happens to be radioactive. As a noble gas, what is breathed into the lung is mostly breathed out again except for a small amount which may be transferred to the blood or decay. The hazard from radon arises from the fact that, when radon in the air decays, its daughter products are not gaseous, and when they are breathed they deposit on the interior surfaces of the lung. Radon has several radioactive daughters. Two of these, polonium-218 (RaA) and polonium-214 (RaC′), emit alpha particles and are the main source of radiation damage when they decay in the lung.

Radon-222 is the immediate radioactive daughter of radium-226, which is a member of the decay chain starting with uranium-238, a very widely distributed element. The uranium-238 decay chain is shown in Table 2-1. Radon-222's half life of 3.8 days makes it possible for it to move considerable distances around the environment from the place at which it was produced.

For a chain of radioactive isotopes headed by a relatively long lived parent, the amount of each chain member will adjust until the sources and losses of each are in balance. When the only source is an original amount of a long-lived parent and the only losses are by the radioactive decay of each member of the chain, the equilibrium reached is one in which the activities of the long-lived parent and each of its shorter lived daughters are equal. The members of the chain are said to be in "secular equilibrium".

Starting from the case in which only the long-lived parent is present, the time it takes to reach an overall equilibrium depends on the rate of loss of each chain member. If one starts from pure radon as the "long-lived" parent, and in the absence of any removal process other than radioactive decay, the time to attain various fractions of the secular equilibrium amount of each of the short-lived daughters is shown in Table 2-2. The members of the chain following polonium-214 are not shown in Table 2-2 because of the long half-life of lead-210. The lead-210 in the air will be removed by deposition onto surfaces before it can radioactively decay. Similarly, the lead-210 deposited in the lung by breathing or by the decay of polonium-214 has a good chance of being removed from the lung by physiological processes before it decays. As a result, the chain is effectively broken just after polonium-214. Overall, as far as induction of lung cancer from radon inhalation is concerned, the only significant members of the decay chain are the daughters before lead-210, i.e. polonium-218 (RaA), bismuth-214 (RaB), polonium-214 (RaC), and lead-214 (RaC′).

F. Radioactivity in the air

A crucial factor in determining the potential hazard of a radon contaminated atmosphere is the relative amounts of radon and of each

TABLE 2-1 *Decay chain from uranium-238 to lead-206 (minor branches, with branching ratios of less than 0.2%, are omitted).*

(1) Isotopic Name	(2) Historical Name	(3) Half-Life	(4) Principal radiations and energies (MeV)
U-238	Uranium I	4.47×10^9 y	α 4.198 (77%) 4.149 (23%)
Th-234	Uranium X_1	24.1 d	β 0.198 (72%)
Pa-234m	Uranium X_2	1.17 min	β 2.29 (98%)
U-234	Uranium II	244 500 y	α 4.773 (72%) 4.721 (27%)
Th-230	Ionium	75,000 y	α 4.688 (76%) 4.621 (23%)
Ra-226	Radium	1 600 y	α 4.785 (94%) 4.602 (6%)
Rn-222	Radon	3.82 d	α 5.490 (100%)
Po-218	Radium A	3.11 min	α 6.003 (100%)
Pb-214	Radium B	26.8 min	β 0.65 (50%)
			γ 0.295 (19%) 0.352 (37%)
Bi-214	Radium C	19.8 min	β up to 3.26 MeV
			γ 0.609 (46%) 1.120 (15%) 1.765 (16%)
Po-214	Radium C'	1.6×10^{-4} s	α 7.687 (100%)
Pb-210	Radium D	22.3 y	β 0.015 (81%)
Bi-210	Radium E	5.01 d	β 1.161 (100%)
Po-210	Radium F	138 d	α 5.297 (100%)
Pb-206	Radium G	stable	none

References: NEA 1985, Lederer 1978, and Walker 1984.

Explanation of columns:

1. The isotopic name is the name currently used in scientific discussions. Standard symbols are used for the elements as follows:

Uranium	U	Radium	Ra	Bismuth	Bi
Palladium	Pa	Radon	Rn	Lead	Pb
Thorium	Th	Polonium	Po		

2. The historical name is the name originally used, before the product was fully identified. These names are still often used for radon daughters, for reasons of convenience in discussion.
3. Half-lives for radioactive decay are specified in years (y), days (d), minutes (min), or seconds (s), as appropriate.
4. The principal radiations are shown for each step of the chain. For alpha particles (α) and gamma rays (γ), the emission energies are shown; for beta particles (β), the indicated energy is the end point (i.e. maximum) energy for the major branch. The relative proportions of the different branches are indicated in parentheses.

TABLE 2-2. *Approximate time to equilibrate radon daughters with radon-222 (in minutes).*

Daughter	Half Life	Time to Indicated Equilibrium (%) 25	50	90	99
Po-218 (RaA)	3 min	1	3	10	20
Pb-214 (RaB)	27 min	16	31	95	180
Bi-214 (RaC)	20 min	36	60	135	230
Po-214 (RaC′)	≪1 sec	36	60	135	230

Reference: Evans 1969.

of its short lived radioactive daughters that are present. The relative amounts may vary anywhere from pure radon to an equilibrium mixture. In this context, it is customary to use the term "equilibrium" to mean the secular equilibrium which exists when the only input is that of radon-222 and the only losses are by radioactive decay.

Departures from this equilibrium occur when additional sources or removal processes alter the relative abundances of radon and its daughters in the air. For indoor radon, the most important of the removal processes are ventilation and the plateout of radon daughters on surfaces within the house. To the extent that these processes are constant over time scales longer than the four hours required to reach equilibrium (see Table 2-2), the concentrations of the chain members down to polonium-214 tend to follow the variations in the sources and losses.

The removal processes reduce the concentrations of radon daughters in the air. Since it is these concentrations which determine the breathing hazard, it is important to know how much the concentrations fall below the equilibrium values.

When one of these radioactive atoms decays, the daughter atom recoils in an ionized state. For decay by alpha-particle emission, the recoil energy is large (about 100,000 eV) while for beta decay, the recoil has a maximum energy of only a few eV. The recoiling ion interacts strongly with the atoms in its neighborhood and is rapidly slowed down and neutralized. The resulting atom, which may have some residual charge, is highly reactive chemically and is likely to react with one of the components of the atmosphere such as oxygen, or trace gases, with consequent change in its subsequent chemical behavior (Busigin 1981). As an ion in air, electrostatic attraction makes it very easy for the daughter to attach to aerosol particles. As a neutral particle, it can also adsorb to aerosols. In either case, the daughter will be deposited with the aerosol in the lung or on room surfaces.

Yeh *et al.* (1976) define an aerosol to be "a system of solid or liquid particles which are (a) dispersed in a gaseous medium, (b) able to remain suspended in the gaseous medium for a long time relative to the time scale of interest, i.e., are relatively stable; and (c) have a high surface area to volume ratio. The geometrical diameters of aerosol particles normally fall within the range between 0.001 and 100 microns." Those daughters which are adsorbed onto the micron and submicron sized atmospheric aerosol are carried with it and are deposited in the lung accordingly. The daughters that are so adsorbed make up the "attached fraction". The remaining daughters in the air occur as a mixture of free ions, oxides, or other chemical forms, usually as very small (angstrom sized) agglomerates with several water molecules. These are called the "unattached fraction" (Raabe 1969).

The daughters in the air occur as a mixture of attached and unattached fractions. When breathed into the lung, each fraction has a different deposition behavior that depends on its particular chemical and physical form. There is a dynamic equilibrium in the air between the two fractions and it is important to know the concentration and composition of each. In consequence, in characterizing the radon daughters in a given environment, it is common to specify the "unattached fraction" of each as one of the parameters to be used in assessing the hazard. The unattached fraction of polonium-218 (RaA) is usually less than 0.10 in indoor environments although higher values are possible in dust-free atmospheres; the unattached fractions of the other daughters are usually much smaller (NCRP 1984).

The decay daughters may plate out on any available surface. For an enclosed space with complex surfaces such as the walls and furniture in a room, plateout and detachment from them are very complicated and data are not readily available. The rate of plateout can be specified in terms of the ratio of the amount deposited onto a surface from the air (in atoms removed per unit area per unit time) to the concentration in the air (in atoms per unit volume). Such a ratio has the dimensions of velocity (length per unit time) and has been dubbed a "deposition velocity".

A recent review of the plateout process (Knutsen 1983) suggested values for the plateout deposition velocity of 0.05 cm/sec for the highly reactive unattached fraction and 0.00075 cm/sec for the fraction which has already attached to the atmospheric aerosol. For a typical room of 12 ft x 10 ft x 7 ft, these deposition velocities correspond to removal rates of 0.064 per minute for the unattached fraction and 0.00096 per minute for the attached fraction.

With these rates, this plateout can compete with radioactive decay and ventilation. Typical air changes range between 0.3 and 1.0 internal volumes per hour which corresponds to removal rates of between 0.005 and 0.0167 per minute. Thus, only the plateout of the unattached fraction is important. In some theoretical analyses plateout is neglected, which gives an over-estimate of the airborne concentrations of the unattached fraction. Since the unattached fraction deposits in the respiratory tract with 100% efficiency, this error leads to an over-estimate of the dose which may be significant except in atmospheres where the unattached fraction is small.

G. Radon daughters: descriptive terminology

1. Working level. The hazard from airborne radon is due to inhalation of its short-lived daughters. The daughters are trapped in the lung and decay there, depositing their alpha particles' energy in the lung tissue.

As a measure of the concentration of short-lived radon daughters in the air in uranium mines, a unit called the "working level" (WL) was introduced. As the name suggests, the WL was taken at that time (but no longer) to represent the maximum concentration of airborne short-lived radon daughters to which uranium miners could safely be exposed. Subsequently, this unit has been applied to the concentration of short-lived airborne radon daughters in buildings occupied by the general population. In most of these locations, the concentrations are well below 1 WL. The definition and use of the WL unit are discussed in the following paragraphs.

As outlined above, for a given concentration of radon in the air, the concentration of its short-lived daughters can vary depending on the removal processes that are acting. The amount of alpha-particle energy that potentially could be deposited in the lung is proportional to the number of short-lived daughter atoms in the air and the amount of alpha-particle kinetic energy each is capable of producing in its decay to long lived lead-210.

An atom of polonium-218 (RaA) decays by the emission of a 6.00 million electron volt (6.00 MeV) alpha particle. Its two immediate progeny, lead-214 (RaB) and bismuth-214 (RaC), decay by relatively low energy beta-particle emission. Beta particles, although they may be energetic, produce very low ionization density and so contribute little to the biological hazard compared to the alpha particles. The grand-daughter, RaC, beta decays to polonium-214 (RaC′) which then decays with a half-life of 164 microseconds, emitting a 7.69-MeV alpha particle. (The RaC′ is essentially always in equilibrium with RaC and RaC is treated as though it were producing the alpha particle.) Although RaB decays by beta emission to RaC and so has no alpha particle of its own, every atom of RaB will eventually produce a 7.69-MeV alpha particle via the decay of its grand-daughter RaC′.

The result of this bookkeeping is that every atom of RaA will produce two alpha particles with a total energy of 13.69 MeV, while every atom of RaB and RaC will produce one 7.69-MeV alpha particle. If there are no removal processes except for radioactive decay, and if the short-lived daughters have reached equilibrium with their radon parent in the air, then the amount of alpha energy which potentially could be released by the short-lived daughters of radon is a maximum.

In the uranium mining industry, an atmosphere contaminated with 100 pCi/ℓ (3.7 dps per liter) of radon gas in equilibrium with its daughters was said to be at 1 WL. For such a situation, the concentration of short-lived daughters is such that each liter of air would potentially produce about 130,000 MeV of alpha-particle kinetic energy.

The working level was *defined* to be that concentration of short-lived radon daughters in any combination which would potentially produce 130,000 MeV of alpha-particle kinetic energy per liter of air. A more exact value for the potential alpha energy generated by 100 pCi/ℓ of radon in equilibrium with its daughters is 128,400 MeV per liter. The definition of the working level therefore means that 1 WL is generated by 101.3 pCi/ℓ of radon in equilibrium with its short-lived daughters. The value of 128,400 MeV/ℓ is used in Table 2-3 so that 100 pCi/ℓ of radon in equilibrium with RaA, RaB and RaC will yield an equilibrium factor of unity. Table 2-3 gives the details of the calculation for the WL′unit.

In the event that removal processes other than radioactive decay are operating, the concentration of the short-lived daughters will be less than the equilibrium amount. In that case, 100 pCi/ℓ of radon will produce fewer than one WL of potential alpha energy.

The number of working levels is obtained by summing over any combination of short-lived daughters. The potential alpha-particle energy concentration is calculated as shown in Table 2-3 and the number of working levels are given by division of this value by the reference equilibrium value of 128,400 MeV per liter. The value obtained by using the definition value of 130,000 MeV/ℓ is not significantly different.

2. Equilibrium factor. The extent to which the short-lived daughters have concentrations less than their values in secular equilibrium with radon can be denoted by the "equilibrium factor", F, where F is defined by the expression:

$$\text{exposure in WL} = 0.01 \times F \times (\text{radon concentration in pCi}/\ell)$$

For example, suppose the concentrations of the short-lived radon daughters relative to their concentrations at secular equilibrium with radon are: $c_A/c_{Rn} = 1/2$, $c_B/c_{Rn} = 1/3$, $c_c/c_{Rn} = 1/6$. Then, applying the method shown in Table 2-3, the equilibrium factor would be $F = 0.29$, and a radon-222 concentration of 100 pCi/ℓ corresponds to 0.29 WL.

3. Unattached fraction . There are several ways of describing the unattached fraction. The specific activity of a radon daughter that is not attached to the atmospheric aerosol can be related to the overall specific activity of that daughter that is actually present (see, for example, Reineking 1985) as:

$$f_i = c_i^u/[c_i^u + c_i^a]$$

where: c_i^a is the specific activity (pCi/ℓ) of RaA ($i = 1$), RaB ($i = 2$), or RaC ($i = 3$) attached to the atmospheric aerosol and c_i^u is the specific activity (pCi/ℓ) of the corresponding Rn daughter which is "free" of or unattached to the atmospheric aerosol. The total radon daughter activity is the sum of attached and unattached activities, $c_i = c_i^a + c_i^u$.

TABLE 2-3. *Working Level: definition and calculations.*

Definition

One Working Level (WL) ≡ any combination of *short-lived* radon daughters in 1 liter of air with the potential of emitting $1.3 \cdot 10^5$ MeV of alpha-particle energy during decay to lead-210. (At secular equilibrium, 1 WL corresponds to 100 pCi/ℓ of radon-222.)

Calculation of potential alpha-particle energy release (PAER)
(assuming 100 pCi/ℓ for each daughter)

$$\mathrm{PAER} = \sum N_i E_i$$

$$N_i = \text{number of atoms per liter of } i^{th} \text{ daughter} = 3.7\, T_{\frac{1}{2}} / \ln 2$$

$$E_i = \text{alpha} - \text{particle energy release for } i^{th} \text{ daughter}$$

daughter nuclide	Half-Life (sec)	N_i	E_i (MeV)	N_iE_i (10^5 MeV)
Po-218 (RaA)	187	996	13.69	0.136
Pb-214 (RaB)	1608	8583	7.69	0.660
Bi-214 (RaC)	1188	6342	7.69	0.488
Po-214 (RaC′)	0.0002	0.001	7.69	$\ll$ 0.001
TOTAL				1.284

Working Level, WL

$$\mathrm{WL} = \frac{13.69 n_A + 7.69(n_B + n_C)}{1.3 \cdot 10^5}$$

where WL = number of working levels and n_A, n_B, and n_C are the concentrations of polonium-218, lead-214 and bismuth-214, respectively, in atoms per liter.

Equilibrium Factor, F

$$\mathrm{WL} = \frac{F \times c_{Rn}}{100}$$

$$F = \frac{0.136 c_A + 0.660 c_B + 0.488 c_C}{1.284 \times c_{Rn}} = \frac{0.106 c_A + 0.514 c_B + 0.380 c_C}{c_{Rn}}$$

where c_{Rn} is the radon-222 concentration in pCi/ℓ, and c_A, c_B and c_C are the polonium-218, lead-214 and bismuth-214 concentrations, respectively, in pCi/ℓ [for example, $n_A = N_1 c_A / c_{Rn}$] and 1.284 is used in place of 1.3 so that F = 1 for 100 pCi/ℓ of Rn in equilibrium with its short-lived daughters.

The specific activity of an unattached radon daughter can be related to the overall equilibrium concentration of that daughter (see, for example, NCRP 1984, Section 6.2) as:

$$f_x = c_x^u / c_{Rn}$$

where: c_x^u is the specific activity (pCi/ℓ) of RaX unattached to the atmospheric aerosol, X stands for A, B, or C, and c_{Rn} is the specific activity (pCi/ℓ) of the radon which defines the equilibrium activity of each of the daughters.

The unattached fraction can be expressed in terms of the fraction of the working level contributed by those daughters which are not attached to the atmospheric aerosol (see, for example, NEA 1983, section 2.4) as:

$$f_p = WL^u / WL$$

$$f_p = \frac{0.106 c_A^u + 0.514 c_B^u + 0.380 c_C^u}{F \times c_{Rn}}$$

where: c_x^u and c_{Rn} are as defined above and F is the equilibrium factor (see Table 2.3).

4. Equilibrium equivalent concentration. The concept of the equilibrium factor defined above gives rise to an alternate specification of the radon daughter concentrations in terms of an equivalent radon concentration. The overall daughter concentrations are here given in terms of the radon concentration that, in equilibrium with its daughters, would generate the number of working levels that are actually present. Thus, if the actual radon concentration is c_{Rn} and the radon daughter concentrations are such that the equilibrium ratio is F, then the "equilibrium equivalent concentration" EEC, is given by:

$$\text{EEC} = F \times c_{Rn}$$

where EEC and c_{Rn} are expressed in pCi/ℓ (or Bq/m^3).

The EEC, similarly to the equilibrium factor, F, does not distinguish between the unattached and the attached fraction. Thus, it is not a strict indicator of the relative hazard of a given effective radon concentration. The dose per WLM from one WL of unattached activity is much larger than the dose per WLM from one WL of attached activity. Thus, it is necessary to know the distribution of the radon daughters between the unattached and the attached activity as well as the EEC. In most indoor environments, the aerosol concentrations are sufficiently high that the contribution of the unattached fraction of RaA to the lung dose is modest and that from the unattached fractions of RaB and RaC is very low.

5. Working level month. Total exposure is dependent on both the intensity of the exposure and its time duration. Assuming that a working year has 2000 working hours, then a working month has 167 working hours if holiday time and overtime are neglected. The working month is usually rounded off to 170 hours.

Breathing an atmosphere contaminated to one working level for one working month constitutes one "working level month" (WLM) of exposure. The incidence of lung cancer among miners exposed to fairly high levels of the short-lived daughters of radon is found to be positively correlated with the number of WLM's they have accumulated.

A miner exposed to 1 WL over a full year will receive, by definition, an annual exposure of 12 WLM. In considering indoor exposures, allowance must be made for the (usually) longer time spent in the indoor environment than miners spend working. For example, if an individual spends 80% of his time in an indoor environment, this is the equivalent of 0.8 x 24 x 365/170 = 41 working months. Thus, an individual exposed to 0.1 WL for 80% of the time receives a total annual exposure of 4.1 WLM (not 1.2 WLM as would be the case for the work-period of a miner).

References: Chapter 2

Busigin A., van der Vooren A. W., Babcock J. C. and Phillips C. R., 1981, "The nature of unattached RaA (^{218}Po) particles," *Health Phys.* **40**, 333.

Evans, R. D., 1969 "Engineers guide to the elementary behavior of radon daughters," *Health Phys.* **17**, 229.

Knutsen E. O., George A. C., Frey J. J. and Koh B. R., 1983, "Radon daughter plateout—II, prediction model," *Health Phys.* **45,** 445.

Lederer C. M. and Shirley V. S., editors, 1978, *Table of Isotopes, Seventh Edition* (New York: Wiley).

Raabe O. G., 1969, "Concerning the interactions that occur between radon decay products and aerosols," *Health Phys.* **17,** 177.

NCRP (National Council on Radiation Protection and Measurements), 1984, "Evaluation of Occupational and Environmental Exposures to Radon and Radon Daughters in the United States," *NCRP Report No. 78* (Bethesda, MD: NCRP).

NEA (Nuclear Energy Agency), OECD, 1983, "Dosimetry Aspects of Exposure to Radon and Thoron Daughter Products," *Report by a Group of Experts* (Paris: OECD).

______, 1985, "Metrology and Monitoring of Radon, Thoron and their Daughter Products," *Report by a Group of Experts* (Paris: OECD).

Reineking A., Becker K. H. and Porstendorfer J., 1985, "Measurements of the unattached fraction of radon daughters in houses," *The Science of the Total Environment* **45**, 261.

Walker F. W., Miller D. G. and Feiner F., 1984, *Chart of Nuclides* (San Jose, CA: General Electric Company).

Yeh H. C., Phalen R. F. and Raabe O. G., 1976, "Factors influencing the deposition of inhaled particles," *Env. Health Perspectives* **15,** 147.

Chapter 3

METHODS FOR DETECTION OF RADON AND RADON DAUGHTERS

Ahmad E. Nevissi

A. General issues in radon detection

1. Scope of chapter. The three naturally occurring isotopes of radon are radon-222 (half-life = 3.8 days), radon-220 (half-life = 55 seconds), and radon-219 (half-life = 4 seconds). As discussed in Chapter 1 (Section E), the activity of radon-219 and its decay products is of minor importance because of its low abundance and short half-life. Further, although most rocks and building materials contain almost equal activities of uranium and thorium and the activity of the radon-220, also known as thoron, in non-porous material is comparable to the activity of the radon-222, the much shorter half-life of thoron causes its concentration in air to be relatively low and therefore usually of secondary interest.

For these reasons, only the measurement methodology for radon-222, or simply "radon", and its daughters will be discussed below and, unless otherwise specified, consideration will not be given to the detection of thoron or to the complications which thoron may introduce into the radon measurements.

The determination of the attached and unattached fractions for the daughter products (see Chapter 2, Section F) will not be discussed in detail here, although calculated radiation doses are dependent upon the assumed values of these fractions. The general approach used for this determination is to pass the air through a fine-mesh screen, or screens, and then onto a filter. The unattached fraction preferentially deposits on the screen and the attached fraction on the filter (NEA 1985). In the absence of such measurements, the dose to the lung is determined on the basis of average values taken to be applicable to the specific environment being considered.

A discussion of thoron detection and of the determination of the attached and unattached fractions is presented in a 1985 report on "Metrology and Monitoring of Radon, Thoron and Their Daughter Products," prepared by a Group of Experts for the Nuclear Energy Agency of the European Organization for Economic Co-operation and Development (NEA 1985). This Report also presents a comprehensive review of techniques for the detection of radon and its daughters.

2. Radon and radon daughters. The medical hazards of radon inhalation stem from the radon daughters in the atmosphere, not from the radon itself (see Chapter 2, Section E). It might be expected therefore that air quality would be uniformly characterized by specifying the radon daughter concentrations, expressed in terms of the number of working levels (WL). However, it is more difficult to measure daughter concentrations than radon concentrations, because the longest-lived daughter has a half-life of only 27 minutes (see Table 2-1). Thus, it is common to measure the radon concentration and, if desired, convert to the number of WL using an assumed value of the equilibrium factor, F (see Table 2-3).

If counting of an air sample begins a few hours after the sample is collected, the radon daughters in the original sample will have virtually all decayed and only the original radon concentration can be determined. However, daughter nuclei are continually produced by radon decay in the sample and the detected counts will include the daughter decays. Thus, there is a distinction between the determination of radon daughter concentrations (which requires counting the sample with very little delay) and the determination of radon concentrations through the counting of radon daughter decays.

3. Problems of low counting rate. A typical indoor air sample has a radon concentration in the neighborhood of 1 pCi/ℓ. This means that in each liter of air, there will be 2.2 decays per minute of radon-222. Even with the further decays of the radon daughters, this implies low counting rates. Each alpha-particle decay of radon-222 is shortly followed by the alpha-particle decays of polonium-218 and polonium-214. Thus, there will be about 7 alpha particles per minute for each picocurie of pure radon. If these were counted with 100% efficiency, with no contributions from extraneous counts, a counting period of about one hour would be needed to yield 400 events. From simple statistical considerations, this means that the radon concentration is determined to within about ten percent (with 95% certainty), if the background rates are low compared to the radon counting rates. For most measurements of outdoor levels and for many measurements indoors, the radon concentration is less than 1 pCi/ℓ, background counts cannot be neglected, and the problems of low counting rates are still greater.

The accuracy of the determination and the speed with which it is obtained is improved if the amount of radon in the sample is larger than the 1 pCi/ℓ assumed in the example above or if the volume of the sample is much larger than one liter. Many of the techniques used for the determination of radon concentrations are designed to achieve higher counting rates by the sampling of a large volume of air. As illustrated below, this can be accomplished by collecting larger initial volumes and later concentrating the radon, by pulling large volumes of air through filters, or by sampling the air for extended time periods.

4. Sampling period. Indoor radon and radon daughter concentrations vary with time, on both a daily cycle and an annual cycle. These variations depend in part on changing behavior of the inhabitants of the house and in part on changing climatic conditions. A choice must be made of the length of the sampling period, based on the type of information desired and the available time. The period can vary from a matter of seconds to continuous periods of up to a year. In the latter case, one obtains fuller exposure information at the expense of obtaining prompt information.

In Sections B and C, methods are described which give the results at a particular time. For all practical purposes, these are instantaneous samplings although they may take up to several hours to carry out. In Section D, continuous counting methods are described, and in Section E a review is given of methods which passively sample over longer times, giving the average concentrations during the sampling time. These are called "integrating methods" because the average is obtained by adding the total number of events and dividing by the time period. Most large-scale programs to survey radon levels in homes use such passive integrating devices.

B. Determination of radon-222 concentrations: instantaneous methods

1. Collection of radon samples from atmosphere. In the "grab sample" technique, the air to be studied is collected in a container and brought back to the laboratory for analysis. Typical containers include plastic bags, metal cans and glass containers. The volumes of the containers are usually between 5 liters and 20 liters. For the measurement of relatively low radon concentrations it is desirable to use a large collection volume and then to concentrate the sample into a smaller volume for counting. Concentration can be accomplished by passing the air through an activated charcoal trap submerged in liquid nitrogen. At this temperature, radon is retained on the charcoal whereas air passes through. The trap is then warmed up and the radon is flushed, with a small amount of helium, into a Lucas counting cell or ionization chamber (see below).

For liquid scintillation counting, the air is bubbled through a chilled liquid scintillator solvent where the radon is dissolved and the air passes through. The solvent is then mixed with a scintillator mixture and the radon in the solvent is counted.

2. Alpha-particle scintillation counting with ZnS: Lucas Cells. One of the oldest and simplest devices for determining radon concentrations uses a technique developed by Lucas (1957). In this technique, the radon gas sample is introduced into a counting cell. The inside wall of the cell is coated with zinc sulfide (ZnS), except one end which is covered with a transparent window for coupling to a photomultiplier tube. When an alpha particle strikes the wall of the cell, a flash of light is emitted from the ZnS coating. The light is detected by the photomultiplier tube and translated into an electrical signal. The efficiency of these cells (i.e., the ratio of the number of electrical signals to the number of alpha particle decays in the cell) is typically 70 to 80% (NEA 1985). Thus, each radon-222 decay leads to the emission of three alpha particles and to the detection, on average, of slightly more than two alpha particles. Background rates in typical Lucas cells are low, about 0.1 or 0.2 counts per minute (cpm).

The alpha particles must reach the ZnS-covered wall to give a signal. Because alpha particles can travel only short distances in air before stopping, the size of the Lucas cell is often limited to a few hundred cubic centimeters making the cells useful only down to radon concentrations of about 1 pCi/ℓ (NEA 1985). However, Cohen and collaborators (1982) have shown, in an extensive effort to optimize Lucas cell performance, that it can be profitable to use cells as large as 3 liters (3,000 cubic centimeters). In this cell, a radon concentration of 0.1 pCi/ℓ gives about 1 cpm with a background of about 0.5 cpm, corresponding to a statistical accuracy of about 11% for a two-hour counting period.

3. Alpha-particle scintillation counting with liquid scintillators. In an alternative scintillation counting technique, a liquid scintillator is used in place of the ZnS, and the radon is mixed into the scintillator. In one application of this approach (Prichard 1983), the radon-bearing air is passed through an organic solvent in which radon is highly soluble at low temperatures. The solvent, with the radon, is then introduced into a vial containing a liquid scintillator. Each radon-222 decay is followed, within several hours, by four additional particles, namely two alpha-particles from the decay of polonium-218 and polonium-214 and two beta-particles from the decay of lead-214 and bismuth-214 (see Table 2-1). The particles are not separately identified, but all five can be detected as events which give relatively large light signals in the phosphor.

This technique permits the use of large air samples together with counting in a cell of small volume. In one example, under conditions where a 10-liter air sample was collected, it was estimated that 37.5% of the radon was extracted and that, on average, 80% of the decays were detected (Prichard 1983). Thus, for a concentration of 0.25 pCi/ℓ, there are 8.3 cpm [10 x 0.25 x 0.037 x 5 x 0.375 x 0.80 x 60 = 8.3]. If the background rate in this detector is 12 cpm, a 100-minute count will determine the radon concentration with a statistical uncertainty of about 5%. This is considerably better than can be obtained in typical small Lucas cell systems, without concentration of the sample, despite the presence of a relatively large background counting rate.

4. Internal ionization chamber counters. Alpha particles from the decay of radon and its daughters can also be detected in ionization chambers. In these counters, an electrical signal is produced without the intermediary of scintillation counting. Ionization counters can be used either to count electrical pulses from individual decay events or to measure currents resulting from the integrated effect of all decays. The Environmental Measurement Laboratory of the Department of Energy uses pulse ionization chambers while the National Bureau of

Standards uses current ionization chambers. In general, ionization chambers are not as widely used as scintillation counters, since ionization chambers are more expensive to construct than Lucas cells and for radon measurements they do not appear to have a major advantage over Lucas cells.

5. Two-filter methods. For measurement of both radon and radon daughter concentrations, the two filter method can be used. In this method, air is passed through the first filter where daughter products are removed. Then the air is passed through a long decay chamber, where daughter products are allowed to grow in and are collected on a second filter. The filters can be counted separately to determine the concentration of radon (from the second filter) and daughter products (from the first filter). The above methods are used for measurement of both grab samples and continuous samples (NEA 1985).

The grab sample measurement is completed in less than 25 minutes at a sensitivity level as low as 0.1 pCi/ℓ for a decay chamber tube volume of 15 liters (Tymen 1978). Several variations of the two filter method are in use that utilize alpha spectrometry, scintillation counting, and a passive detector for measurement of radon and its daughter products.

C. Determination of radon daughter concentrations: instantaneous methods

1. General. Unlike radon, radon daughters deposit readily on dust particles or other surfaces. By drawing air through a filter, the radon daughters can be collected with high efficiency. Most methods of determining radon daughter concentrations are based on drawing known volumes of air through a filter and counting the deposited activity. Other methods utilize electrostatic deposition onto a detector. In all cases, counting must begin shortly after the sample is collected because the daughters all have short half lives (the longest is 27 minutes).

In order to determine the daughter concentrations, expressed as the number of working levels, it is necessary to know the individual concentrations of RaA (polonium-218), RaB (lead-214) and RaC (bismuth-214) (see Table 2-3). The half-lives for the decays of these daughters are 3.11 minutes, 26.8 minutes, and 19.8 minutes, respectively. The concentration of RaC′ (polonium-214) is not relevant, because its half-life is so small (< 0.0002 sec) that only a negligible number of RaC′ nuclei are present on the filter at any instant.

The three concentrations could be determined simultaneously if it were practical to detect all of the emitted alpha-particle, beta-particle, and gamma-ray groups, identifying each as to type of particle and energy. However, such comprehensive counting faces practical difficulties and is rarely, if ever, attempted. Instead, alternative expedients are adopted, as discussed in the succeeding sections. Many of them rely on counting of alpha particles because they are the easiest to detect with simple equipment.

2. Alpha particle scintillation counting: single time period. A very simple approach is to count the alpha-particle radioactivity deposited on the filter, using a scintillation counter with a ZnS phosphor. The filter containing radon daughters can be covered with a thin sheet of ZnS phosphor (commercially available) and then placed on the photomultiplier tube for counting, or the ZnS can be left on the photomultiplier tubes. This is the method originally developed by Kusnetz (1956). If counting is continued for a protracted period, each RaA atom on the filter at the start of counting will give rise to a 6.00-MeV and a 7.69-MeV alpha particle and each RaB and RaC atom will give rise to a 7.69-MeV alpha particle. In this simple version, no attempt is made to distinguish between the alpha-particle energies. Were the relative daughter concentrations known for the original air sample, the single count would establish the absolute total daughter concentration and with it the individual daughter concentrations. However, in general the relative daughter concentrations are not known and therefore this method yields only a rough measure of the number of WL. One version of Kusnetz method uses a 5 minute sampling period at a flow rate of 2 liters per minute, allows the sample to decay for 40 minutes, and then counts the total number of alpha particles for 2 minutes (NEA 1985). This method, with its short counting time, was developed for radon monitoring in uranium mines, which have substantially higher radon levels than most indoor environments.

3. Alpha-particle scintillation counting: several time periods. In principle, the three radon daughter concentrations can be individually established if counts are taken during three or more successive time periods. Because the half-lives of the three daughter nuclei are different, the relative number of counts in three time periods determines their individual initial concentrations. This method was first used by Tsivoglou (1953) and has been modified and refined by subsequent investigators. A summary of sampling and counting times for different one- two-, and three-count methods is given in NEA 1985. It is also possible, especially with the aid of computer-based data analysis systems, to count for many short periods.

4. Alpha-particle spectroscopy. As an alternative to counting for three or more time periods, it is possible to obtain the daughter concentrations by distinguishing in the detecting system between the 6.00-MeV alpha-particle group from RaA and the 7.69-MeV alpha-particle group from RaC′ and separately counting the two groups during two time periods. This, in fact, gives redundant information, with four measured numbers being available to determine three unknowns. An automatic method for measuring radon daughters by alpha spectroscopy using silicon surface-barrier detectors is described by Cliff (1978).

Alpha-particle spectroscopy can also be done with one time period. The number of 6.00-MeV alpha particles determines the initial concentration of RaA. The concentrations of RaB and RaC are then determined from the number of 7.69-MeV alpha particles, together with an estimate of the relative concentrations of RaB and RaC. This method has been termed the "rapid spectroscopic technique" (Revzan 1983).

5. Combined alpha-particle and beta-particle spectroscopy. The counting of beta-particles from RaB and RaC in a plastic scintillator along with alpha-particle spectroscopy with a surface-barrier detector determines the three daughter concentrations during a single time period (NEA 1985). This method is clearly advantageous in terms of speed of measurement, but more complex in terms of the equipment required.

D. Continuous counting methods

Continuous monitoring of the radon concentration (CRM) can be done with a scintillation chamber through which filtered air is drawn. The chamber is coupled with a photomultiplier tube and associated electronics. Different integration intervals, usually 180 minutes, are used for analysis of CRM data. The filtered air is drawn continuously through the chamber and the scintillations from the alpha particles emitted by radon and radon daughters that decay within the chamber are seen by the photomultiplier. At a constant flow of air, the scaler readout should be proportional to the radon concentration, assuming there is no dead space in the cell. An automated system for simultaneous measurement of radon and air exchange rates in homes is described by Nazaroff *et al.* (1983).

For continuous measurement of relatively high concentrations of radon, such as in well water or mine tailings, an open scintillation cell may be used directly. The scintillation cell is coupled to a photomultiplier at one end. The open end, usually connected to an extension tube, is inserted directly into the test medium. The phototube output is monitored continuously, as the response varies with changing radon concentration (Noguchi 1977).

E. Integrating techniques for measuring radon and radon daughters

1. General characteristics of integrating techniques. The methods discussed in Sections B and C, involve prompt counting of the emitted particles, predominantly alpha particles. In this Section, a number of integrating methods are described. In most of these, alpha particles leave a record on a radiation-sensitive device and the device is counted at a later time. Commonly, this method is used for long periods of monitoring, usually for several months, and the recorded events constitute an integration of the alpha-particle decays which have occurred over the full monitoring period. One intermediate case is also discussed, that of adsorption of radon on charcoal, where integration periods of several days are used and the activity is counted before the decay of radon-222, with a half-life of 3.8 days, has progressed too far.

A concise review of integrating techniques, other than charcoal adsorption, is included in the *Health Physics* review volume on indoor radon (Nyberg 1983). Implementation of the charcoal adsorption technique has been discussed in a series of papers by Cohen and collaborators (1983, 1986a).

2. Etched track detectors. The etched-track detector method is perhaps the most widely used method for time-integrated monitoring, and is commercially available under the trade-name Track Etch, marketed by the Terradex Corporation. This general method was originally developed for detecting heavy particles in nuclear physics experiments and from cosmic rays (Fleischer 1965), and was subsequently applied to the specific problems of radon detection (see, e.g., Fleischer 1980; Alter 1981).

The method exploits the fact that a heavy atomic particle, such as an alpha particle, leaves a microscopic track of damage when passing through certain plastics. The tracks are enlarged by chemical etching and counted under a microscope. The number of tracks per unit area is proportional to the radon concentration. Most commonly, radon daughters are excluded using a filtering membrane which admits radon only. If a 17 square-millimeter detector area is scanned, a 30-day exposure at a concentration of 1 piC/ℓ gives about 20 tracks (Nyberg 1983), corresponding to a statistical uncertainty of about 25%.

3. Thermoluminescent detectors. Ionizing radiation can cause atomic or molecular disturbances in some materials such that the material subsequently emits light when heated. In many cases the material is very stable, and this thermoluminescent property remains intact even if many years elapse between the original disturbance and the heating. Thermoluminescent detectors (TLD) for ionizing radiation are based on this property.

The TLD chip is exposed to the alpha and beta particles and the gamma rays that are emitted by radon and its decay products. Usually, a second TLD chip is used for monitoring the gamma ray background radiation.

In one system based on this method, designated as the radon progeny integrating sampling unit (RPISU), air is drawn through a filter, which catches the radon daughters. The dose integrating element is a TLD chip which is located near the surface of the filter (Nyberg 1983). For sampling carried on over a one-week period with air drawn at a rate of 1 liter per minute, it is possible to measure levels as low as 0.0001 WL, corresponding to 0.02 pCi/ℓ if the equilibrium factor is F = 0.5 (Nyberg 1983).

Another system, the so-called passive environmental radon monitor (PERM), uses electrostatic attraction to collect radon daughters from the decay of radon, following diffusion of the radon through a barrier. The dose integrating element is a lithium fluoride chip. The system was originally designed by George and Breslin (1977) and a modified version of it, available commercially, is used by the EPA. The lower detection limit of the modified version is about 0.3 pCi/ℓ for a one week sampling period (Nyberg 1983).

4. Charcoal adsorption detectors. Radon, like a number of other gases, can be adsorbed on charcoal. This property has been utilized to develop a practical detection system in which radon gas is accumulated in a bed of charcoal and the gamma-ray activity from the decay of radon daughters is counted. The detector is useful for integrating radon concentrations over a period of up to about one week. Much longer integration periods cannot be attained due to the 3.8-day half-life of radon.

One advantage of such devices is the simplicity of construction. In the configuration described by Cohen and Nason (1986), a 1.5-centimeter deep bed of charcoal is placed in a commercial metal "ointment can" which has a 3-inch diameter and a 1-inch height. A 3/4-inch diameter hole is drilled in the top of the can and covered with a fine-mesh screen. This allows radon to enter by diffusion, with little convection. A dessicant is added to remove water vapor.

Before the detector is sent to the measurement site, it is baked out and the hole is covered with tape. At the site, the tape is removed and the detector is exposed to the ambient atmosphere for about one week, after which the tape is replaced. The detector is then returned to the laboratory and counted. Counting must begin within several days after the end of the exposure, because of the radon-222 decay.

Counting is carried out with a NaI gamma-ray detection system, set to detect the 295-keV and 352-keV gamma rays from the decay of lead-214 (RaB) and the 609-keV gamma ray from the decay of bismuth-214 (RaC). In the configuration used, a detector exposed for one week to 1 pCi/ℓ of radon and counted for 30 minutes, starting 3.8 days after the end of the exposure, will give a total of about 260 counts from the gamma rays of interest. During the same time interval, there will be 1600 background counts in the energy region of interest, giving a result with a statistical standard deviation of 17%.

Because the detectors are passive and one need only place them in the location to be studied, both this technique and the Track-Etch technique described above permit programs of large-scale radon testing without requiring large number of field workers and without inconveniencing the residents of the home being tested. Cohen has compared the advantages of these two types of detectors (1986). As seen above, better statistical accuracy is achieved for a one-week exposure of a typical charcoal adsorption detector than for a one-month exposure of an etched-track detector, even assuming a scanning area of 17 square-millimeters. Usually the scanning area is considerably less. Cohen also reports that charcoal adsorption detectors have better reproducibility. For these reasons, they are usually preferable when prompt results are wanted and an integration period of more than a few days is not desired. The etched-track detector is superior when long-term averaging is desired and repeated short-term sampling is inconvenient.

F. Radon measurement in water

1. Liquid scintillation counters. As discussed in Chapter 4, radon concentrations in water are typically much higher than the concentrations in air. When the radon concentration in water is sufficiently high (> 1000 pCi/ℓ), as is often the case with well water, direct liquid scintillation counting is a rapid and practical method. The water sample can be mixed with the counting material and counted by the conventional liquid scintillation counters used for radon in air samples. This method lends itself to large scale counting with automation. For example, Hess and collaborators (1982, 1985) have measured radon in 2000 well-water samples.

2. Gas extraction. A more sensitive method of detecting radon in water, suitable at lower concentrations, is to extract the radon as a gas and count the emitted alpha particles in a ZnS scintillation cell. Helium is bubbled through the water, stripping the radon. The mixture of gases is then passed through a cold trap, for example activated charcoal at liquid nitrogen temperature, that traps radon while the helium passes through. The trap is then warmed and radon is transferred into a Lucas counting cell by stripping with a small amount of helium. This technique has been developed and refined by Smethie (1979) for the measurement of small amounts of radon in sea water.

3. Direct gamma-ray counting. When the radon concentration in water is relatively high (> 500 pCi/ℓ), it is possible to determine the radon concentration by counting gamma rays from radon daughter decay using standard gamma-ray spectroscopy techniques with a Ge(Li) detector. The original radon concentration can be distinguished from the radium-226 concentration by repeating the count after 30 days, at which time the original radon will have virtually all decayed and the only remaining radon is that in secular equilibrium with radium-226.

G. Radon intercalibration exercise

In order to establish the extent to which there is consistency between radon measurements made at different institutions, the Environmental Measurements Laboratory (EML) of the Department of

Energy has conducted a series of intercalibration exercises since 1981 (Fisenne 1983, 1985, George 1985). In these exercises, a set of detectors to be tested are sent to the EML, exposed to a standard concentration of radon in a large calibration chamber, and returned to the participating institutions for counting. The results are expressed as the ratio of the mean concentration measured at a given institution to the reference concentration as measured by the EML. The departure of this ratio from unity for a given detector is an indication of the error or uncertainty in measurements made with that detector.

References: Chapter 3

Alter H. W. and Fleischer R. L., 1981, "Passive integrating radon monitor for environmental monitoring," *Health Phys.* **40**, 693.

Cliff K., 1978, "Low concentration radon daughter measurements," *Phys. Med. Biol.* **23**, 55.

Cohen B. L., El Ganayani M. and Cohen E. S., 1982, "Large scintillation cells for high sensistivity radon concentration measurements," *Nucl. Instr. and Meth.* **212**, 403.

Cohen B. L. and Cohen E. S., 1983, "Theory and practice of radon monitoring with charcoal adsorption," *Health Phys.* **45**, 501.

Cohen B. L. and Nason R., 1986, "A diffusion barrier charcoal adsorption collector for measuring Rn concentrations in indoor air," *Health Phys.* **50**, 457.

Cohen B. L., 1986, "Comparison of nuclear track and diffusion barrier charcoal adsorption methods for measurement of ^{222}Rn levels in indoor air," *Health Phys.* **50**, 828.

Fisenne I. M., George, A. and McGahan M., 1983, "Radon measurement intercomparisons," *Health Phys.* **45**, 553.

Fisenne I. M., George A. C. and Keller H., 1985, *The July 1984 and February 1985 Radon Intercomparison,* U.S. Department of Energy, New York, USDOE EML-445.

Fleischer R. L., Price P. B. and Walker R. M., 1965 "Solid state track detectors: application to nuclear science and geophysics," *Annual Rev. of Nuclear Sci.* **15**, 1.

Fleischer R. L., Giard W. R., Mogro-Campero A., Turner L. G., Alter H. W. and Gingrich J. E., 1980, "Dosimetry of environmental radon: methods and theory for low-dose, integrated measurements," *Health Phys.* **39**, 957.

George A. C. and Breslin A. J., 1977, "Measurement of Environmental Radon with Integrating instruments," *Workshop on Methods for Measuring Radiation in and around Uranium Mills* Vol. 3, No. 9.

George A. C., Hinchliffe L., Fisenne I. M. and Knutson E. O., 1985, *Intercomparison and Intercalibration of Passive Radon Detectors in North America,* U.S. Department of Energy, New York, USDOE EML-442.

Hess C. T., Weiffenback S. A., Norton, S. A., Brutsaert W. F. and Hess A. L., 1982, "Radon-222 in potable water supplies in Maine:

The geology, hydrology, physics and health effects," in: *Natural Radiation Environment* (edited by K. G. Vohra et al.), pp. 216-220 (New Delhi: Wiley Eastern).

Hess C. T., Korsah J. K. and Einloth C. J., 1985, *Radon in Houses Due to Radon in Potable Water,* Land and Water Resources Center, Univ. of Maine at Orono, ME 04469, G846-04, G910-03.

Kusnetz H. L., 1956, "Radon daughters in mine atmospheres: a field method for determining concentrations," *Am. Ind. Hyg. Assoc. J. Quart.* **17**, 85.

Lucas H. F., 1957, "Improved low level alpha scintilation counter for radon," *Rev. Sci. Instr.* **28**, 68.

Nazaroff W. W., Offermann F. J. and Robb A. W., 1983, "Automated system for measuring air-exchange rate and radon concentration in houses," *Health Phys.* **45**, 525.

NEA (Nuclear Energy Agency, OECD), 1985, "Metrology and monitoring of radon, thoron and their daughter products," *Report by a Group of Experts* (Paris: OECD).

Noguchi M. and Wakita H., 1977, "A method for continuous measurement of radon in ground water for earthquake prediction," *J. Geophys. Res.* **82**, 1353.

Nyberg P. C. and Bernhardt D. E., 1983, "Measurement of time-integrated radon concentration in residences," *Health Phys.* **45**, 539.

Prichard H. M., 1983, "A solvent extraction technique for the measurement of Rn-222 at ambient air concentrations," *Health Phys.* **45**, 493.

Revzan K. L. and Nazaroff W. W., 1983, "A rapid spectroscopic technique for determining the potential alpha-energy concentration of radon decay products," *Health Phys.* **45**, 509.

Smethie W. M., 1979, *An investigation of vertical mixing rates in fjords using naturally occurring radon-222 and salinity as tracers,* Ph.D. thesis, Univ. of Washington, Seattle, WA 98190.

Tsivoglou E. C., Ayer H. E. and Holaday D. A., 1953, "Occurrence of non-equilibrium atmospheric mixtures of radon and its daughters," *Nucleonics* **11**, 40.

Tymen G., 1978, *Repartition granulométrique de l'aérosol naturel et des particules radioactives issues du radon, en atmosphére maritime et urbaine peu polluée,* Thése de doctorat d'etat, Brest, France.

Chapter 4

RADON SOURCES AND LEVELS IN THE OUTSIDE ENVIRONMENT

Ahmad E. Nevissi and David Bodansky

A. Sources of radon

1. Soil as a radon source. Overall, as a global average, at least 80% of the radon emitted into the atmosphere comes from the top layers of ground (NCRP 1984b). The radon emanation is associated with the presence in the ground of radium and its ultimate precursor uranium. Although these elements occur in virtually all types of rock and soil, their amounts vary with the specific site and geological material. Uranium concentrations are commonly expressed in parts per million by weight (ppm) or in terms of the "specific activity" expressed in picocuries of uranium-238 per gram of material (pCi/g). These units are related by the conversion factor for uranium-238: 1 pCi/g = 3.0 ppm.

To consider the uranium concentrations in some typical rocks, granite, which is relatively rich in uranium, has an average concentration of about 1.6 pCi/g and basalt, which is relatively uranium-poor, has an average concentration of about 0.3 pCi/g (NCRP 1976). The average value for rock in the earth's crust is about 1 pCi/g. Soils average about 0.7 pCi/g. Much higher concentrations are possible, for example minable ores have uranium contents of up to several tens of thousand ppm and there is a broad intermediate range. However, for estimating the general outdoor level of radon, the average concentrations are of greatest relevance.

The specific activities of radium-226, the immediate precursor of radon, and of uranium-238 are equal if uranium and its daughters have remained together for a long time period, so that secular equilibrium holds. Although differences in the geochemical behavior of radium and uranium in the earth's crust cause local departures from secular equilibrium, the average specific activities are nonetheless approximately equal. A rough average radium concentration in the soil is 0.8 pCi/g (NCRP 1976).

Each decay of a radium-226 nucleus produces a radon-222 atom. If the radon is produced close to the earth's surface, it can escape rather than remain trapped in the ground. The average measured emanation rate of radon from soil is about 0.5 pCi per square-meter per second, corresponding to the escape of 8800 atoms per square-meter per second (e.g., NCRP 1984b). Such a number can be understood in terms of a simple calculation. Roughly speaking, about 10% of the radon produced in the top meter of soil escapes (NCRP 1984b). Consider a cube which is 1 meter in each dimension. Using rounded numbers, if the average density of the soil is 2.0 grams per cubic-centimeter (cc) and the average radium-226 concentration is 1.0 pCi/g, the cube will contain 2 million grams of soil and 2 x 10^{-6} Ci of radium-226. This corresponds to the production of 7.4 x 10^4 radon atoms per cubic-meter per second and the escape of 7400 atoms per square-meter per second, in rough correspondence to the average measured value. In alternative units, the figure of 0.5 pCi per square-meter per second corresponds to the emission of 16 Ci of radon per square-kilometer per year.

Thus, approximately 2.4 x 10^9 Ci of radon per year are emitted from the land areas of the earth (1.5 x 10^8 square-kilometers). To consider a more local scale, and assuming the same average emanation rate, approximately 120 million Ci of radon are emitted annually in an area equal to that of the United States (7.7 x 10^6 square-kilometers excluding Alaska) and about 2.5 million Ci of radon per year are emitted in a land area equal to that of the State of Washington (1.7×10^5 square-kilometers).

These numbers are based on average values. The radon emanation rate varies from place to place due to differences in radium concentration and soil permeability. For example, the Bonneville Power Administration reports emanation rates ranging from 0.1 to 2.5 pCi per square-meter per second in regions surveyed in the Pacific Northwest (BPA 1984). The water content of the soil also effects the emanation rate, enhancing it at low water content and diminishing it for moderate or high water content. In addition, emanation is reduced if the ground is covered with snow or if the atmospheric pressure is relatively high. Thus, there will be variations in radon emission rate with time, as well as with location.

2. Ground water as a source of radon. The second most important contributor to outdoor radon is emanation from ground water sources. Ground water in contact with crustal rock penetrates into the pores and voids present in rocks and soils and dissolves radon that emanates into these spaces following radium-226 decay. Radon is quite soluble in water. In equilibrium with radon gas at a partial pressure of one atmosphere, the solubilities range from 51 cc of radon per 100 cc of water at 0 °C to 13 cc of radon per 100 cc of water at 50 °C (Chemical Rubber Publishing 1978). Even the lower figure corresponds to 160,000 Ci per liter of water. Thus, the concentration of radon in underground water is not solubility limited even at high temperatures.

In above-ground air, radon partial pressures are infinitesimal compared to one atmosphere, but sufficiently high radon partial pressures exist in underground rock pores to permit very high concentrations of radon in the water. When this water reaches the surface, most of the radon is released into the atmosphere, due to the reduction in the radon partial pressure, but the remaining small fraction can still correspond to a very large amount of radioactivity. This is one reason why very high concentrations of radon can be found in some hot springs and well waters, although even the highest concentrations fall something like a trillion times below the 160,000 Ci/ℓ figure cited above.

The concentration of radon in ground or well water depends strongly on the character of the host rock. Hess and colleagues (1982) measured radon contents of some two thousand samples from public and privately drilled wells in Maine and reported values ranging from 200 to 50,000 pCi/ℓ of water. The average water radon values obtained from 10 different granite bodies ranged from 1500 to 39,000 pCi/ℓ. As an indication of the global range, one summary (NCRP 1984a) cites studies with average values for radon concentration ranging from a low of 500 pCi/ℓ (for 172 wells in Sweden) to 170,000 pCi/ℓ (for 11 wells in Nova Scotia). This set of studies may be biased towards regions of high radon concentration, and levels of several hundred to several thousand pCi/ℓ are perhaps more typical of ground water (NCRP 1975).

Because of the solubility of radon in water, very high levels of radon can occur even at the boiling point of water or higher temperatures, as reached in some hot springs. These spas have been used traditionally for therapeutic purposes, sometimes specifically with radon as the intended therapeutic agent. In studies of two Austrian spas (Pohl-Ruhling 1982), air activities in the periphery of the spa were 0.1 to 1.5 pCi/ℓ outdoors and 1 to 5 pCi/ℓ indoors, while in the bathhouse itself they reached as high as 3000 pCi/ℓ. In studies of Polish spas (Chruscielewski 1983), the radon content of the water was 3000 to 30,000 pCi/ℓ and in the air of one "inhalatorium" the concentration reached 13,400 pCi/ℓ. Workers in one of the spas suffered exposures of up to 300 WLM per year in 1976, although efforts have been made subsequently to reduce exposures in the Polish spas. It may be noted that 300 WLM is 75 times the recommended exposure limit for United States uranium miners (see Chapter 8, Section B-2).

However, despite the very high levels which can be found in some locations even apart from spas, ground water is responsible for only a small fraction of the radon emitted into the earth's atmosphere. The upper limit for the global contribution is estimated to be 5×10^8 Ci per year, or about one-fifth the amount released from the soil (NCRP 1984a).

3. Other sources of atmospheric radon. The concentration of uranium and radium in sea water is much less than in soils and rocks. Typical values are about 0.9 pCi/ℓ (0.0009 pCi/g) (Joseph 1971). Due to this low concentration and to the high solubility of radon in water, little radon is released from ocean surface waters into the atmosphere. Thus, virtually the entire contribution to the radon in the atmosphere comes from soil and ground water. A compilation of the global sources is given in Table 4-1. It is seen that the oceans, although they have over twice the area of land, contribute only about 1% to the total radon emission into the atmosphere. Uranium mill tailings also make

TABLE 4-1. *Sources of global atmospheric radon.*

Source	Input to Atmosphere (million Ci per year)
Emanation from soil	2000
Ground water (potential)	500
Emanation from oceans	30
Phosphate residues	3
Uranium mill tailings	2
Coal residues	0.02
Natural gas	0.01
Coal combustion	0.001

Reference: NCRP 1984b, Table 3.1.

only a very small contribution, but these will be discussed separately in Section C-2 in view of the special interest which people have had in this source.

B. Radon levels in the outdoor atmosphere

Radon enters the atmosphere from the ground. In consequence, the radon concentrations in air decrease with increasing altitude. In one summary of several different studies (Gesell 1983), it was found that the radon concentration dropped by about a factor of two in the first meter, at roughly the rate of another factor of two in the next 100 m, and roughly another factor of two in the next kilometer. In very round numbers, according to this picture, the concentration falls about a factor of ten in the first kilometer.

For this reason, it is necessary to specify the altitude when talking of outdoor radon concentrations. Comprehensive surveys of actual outdoor concentrations are not readily available. In the summary cited above (Gesell 1983), results of a number of surveys were presented, corrected to correspond to a height of 1-2 meters above the ground. For four regions with no unusual geological features, the concentrations were in the neighborhood of 0.25 pCi/ℓ; in Alaska they were as low as 0.01 pCi/ℓ, presumably due to snow or ice cover, and in an uranium-rich area in Colorado about 0.75 pCi/ℓ. For use in estimating typical outdoor levels, the authors of NCRP Report No. 78 (NCRP 1984b) recently adopted a value of 0.18 pCi/ℓ, which was the geometric mean found for a set of measurements in New York and New Jersey. Thus, it would appear that a typical outdoor level is in the neighborhood of 0.2 pCi/ℓ, but there are wide variations about this average.

An outdoor air concentration of this magnitude can be understood from a rough estimate, based on a radically simplified model. Because most of the radon is in the lower kilometer, consider a column of air 1 kilometer in height and 1 square-meter in cross-sectional area (total volume = 1.0 x 10^6 liters). As seen above (Section A-1), about 8800 atoms of radon enter this volume per second. Treating this column as isolated from the rest of the atmosphere, in the steady state 8800 atoms must therefore decay per second, corresponding to 240,000 pCi. The average concentration in the 10^6-liter column is then about 0.2 pCi/ℓ. (This crude calculation gives a nearly correct result in part because the errors tend to cancel. In particular, neglecting the fact that the oceans contribute little to the global total raises the calculated result and neglecting the fact that the concentrations are higher near the surface than for the 1-kilometer average lowers it.)

C. Radon from man's technical activities

1. Phosphate mining. Phosphate ores have relatively high concentrations of uranium and radium, and high radon levels are encountered in regions of phosphate mining. Typically, the maximum radium-226 concentration in a phosphate ore deposit is about 40 pCi/g (Roessler 1983), or 50 times greater than the average concentration in ordinary soil. This can lead to high radon emanation rates if the deposit is close to the surface or when the products of mining are brought to the surface.

The problem is somewhat localized to Florida, where most of the U.S. phosphate mining is done. It has been studied by the EPA (Guimond 1979) as well as by groups sponsored by industry and State organizations in Florida. The primary concern is over the resulting high indoor radon levels, especially for houses which have been built on land reclaimed after strip mining, with the mining residues used as land fill. Further discussion of this problem is presented briefly in Chapter 5.

2. Uranium mill tailings. Uranium mining represents a problem somewhat analogous to phosphate mining, but with residues richer in radioactive materials. The residue remaining after uranium is extracted from uranium ore constitutes the mill tailings. This residue contains virtually all the mass of the ore and much of the radioactivity. In particular, the uranium-238 daughters, thorium-230 with a half-life of 75,000 years and radium-226 with a half-life of 1600 years, largely remain in the tailings. In most cases, they have been brought into more direct contact with the environment than was the case before the ore was mined.

The main hazards from the tailings comes from the escape of radon to the atmosphere and from leaching of radium into surface water, assuming that the tailings are somewhat stabilized so that the solid elements are not carried into the atmosphere to any large extent. The scale of the problem can be estimated by considering the amounts of uranium mined. For each year of operation of a large (1000 megawatt) nuclear reactor, about 170 tonnes of uranium are used, with total uranium-238 and radium-226 activities of about 60

Curies. If the ore is 0.1% uranium, the mass which goes into the tailings is about 170,000 tonnes, and the radium-226 concentration is about 350 pCi/g. For a large-scale nuclear economy, say one which has operated 400 reactors for 30 years, (we now have about 100 operating power reactors in the United States) the tailings would have a mass of about 2 x 10^9 tonnes and an activity of about 700,000 Ci in each of the daughters.

Escape of radon from the mill tailing depends upon the thickness of the tailings and the nature of any cover. If 20% of the radon produced from the decay of the radium-226 escapes from the tailings (NCRP 1984b), this would correspond to an emanation rate of about 9 million Curies of radon per year. This can be compared, for example, to the total natural radon emanation from the land area of the United States of about 120 million Curies per year (see Section A-1). Thus the tailings, in this picture, would increase the average radon emission from the surface (for the U.S.) by about 8%. If distributed uniformly in the atmosphere over the U.S. this would correspond to an increase in the outdoor radon level of about 0.02 pCi/ℓ. Average indoor concentrations would increase by the same absolute amount, corresponding to an increase of about 1% or 2% over the level without mill tailings.

Of course, a model of uniform distribution over the United States is unrealistic. Winds can spread the radon over the oceans, decreasing the average U.S. concentration, but also there will be higher concentrations in the vicinity of the tailings sites. In a near reductio ad absurdum of enhancement, mill tailings were used in Grand Junction Colorado as fill under buildings. The resulting radon levels are discussed briefly in Chapter 5.

The use of mill tailings for fill in building construction was an avoidable mistake, and remedial action has been taken for the houses involved. The most substantial future hazard comes from the radon escaping into the atmosphere from the tailing piles. A typical radon flux from tailings is 300 pCi per square-meter per second (Rogers 1980), but considerably higher fluxes may also occur. This average is about 600 times the average emanation from soil, leading to the possibility of high atmospheric radon concentrations in the vicinity of the tailings site.

In the absence of the protective measures to reduce the radon emanation, essentially the covering of the tailings, there could be a substantial long-term health impact from the tailings. Robert Pohl (1975, 1982) has argued that the effects of the tailings should be calculated over the full decay time of the thorium-230. He estimated that over this time (roughly the 110,000 year "meanlife" of thorium-230) there would be 400 deaths per gigawatt-year of electricity. Thus, for example, for the program envisaged above, where 400 reactors operate for 30 years, mill tailings would be responsible for some 4 or 5 million deaths, at a rate of about 40 deaths per year. A slightly lower estimate has been given by Cohen (1982), but the qualitative result is not greatly changed.

These numbers assume that the standard estimates of the number of lung cancers per WLM of radon exposure are accurate, despite the large uncertainties in such estimates, especially at the low dose rates involved here (see Chapter 8). They also assume no change in the size of the U.S. population and no improvement in cancer therapy. Thus, they can be considered more as numerical exercises which indicate the scale of the potential impact rather than as reliable estimates of actual future damage. This impact can be viewed as large if the possible several million fatalities over 100,000 years are emphasized and, conversely, it can be viewed as small in that the tailings would add less than one percent to the roughly 10,000 cancers fatalities per year attributed to "natural" indoor radon.

These different perspectives notwithstanding, it is now national policy to treat the tailings problem as an important one, and stringent standards have been imposed to reduce the radon emissions from tailings piles. In particular, standards established by the Environmental Protection Agency in 1983 require that average radon releases from tailing sites be less than 20 pCi per square-meter per second (EPA 1983). The emanation of radon can be controlled by placing covering material over the tailings, trapping most of the radon before it escapes. If the covering is typical Western soil, the radon emanation is reduced about a factor of 2.7 in one meter. Thus, several meters of covering soil would reduce the radon emanation sufficiently to meet the presently mandated limits. Tests have also been made of alternative sealing methods, such as using thin layers of asphalt emulsions, ranging from several millimeters to several inches in thickness (Hartley 1980).

The EPA estimates that "the added lifetime risk of fatal lung cancer for someone living 600 meters from the center of a model pile is 1 in 1000 due to radon from a tailings pile emitting radon at a level of 20 pCi/m^{2}s" (EPA 1983:45937). As will be seen in Chapter 9, this added risk for the close neighbors of a uranium tailings pile is considerately less than the risk the average person in the United States already has of suffering lung cancer due to indoor radon.

Of course, it is a question as to whether the protective measures will be successful over a long period of time. In addition to the technical uncertainties, it is not clear when the mandated protection measure will be enforced. The Uranium Mill Tailings Radiation Control Act was passed in 1978, but, as reported in *Science* (Crawford 1985), implementation of this Act has been "...delayed in part by jurisdictional disputes between the Nuclear Regulatory Commission and the Environmental Protection Agency. But what appears to have been most disruptive is repeated congressional meddling and unresolved litigation brought by industry and environmentalists." Estimated costs of cleanup are from \$2 billion to \$4.4 billion, but the ailing uranium industry is in no position to meet these costs. Aside from uncertainty as to who will shoulder the costs, there is a legal dispute with industry which considers the requirements to be too costly and too stringent and with some environmentalists who believe they are too lenient. Thus, there appears to be uncertainty as to the final form the requirements will take, as well as in the long-term technical success with which they will be implemented. Final implementation of mill tailings stabilization can be delayed until the mills are finally closed (Crawford 1985), and thus the crucial decisions can legally be deferred for a time.

Whatever the outcome of the final codification and implementation of tailings protection, there is no plausible scenario in which the total radon exposure from mill tailings will approach in magnitude the total population exposures from indoor radon. Even with a substantial nuclear power program (12,000 reactor-years) and without remedial measures, the tailings would add less than only about one or two percent to the overall radon exposures. (If nuclear power continues to be used much beyond this 12,000 reactor-years, breeder reactors will probably be employed and much less uranium mining will be required for a given output of energy.) However, there could be elevated local exposures from the tailings. The fact that nature also provides very high radon levels in some localities would not satisfactorily answer the concerns of those living near tailings sites, and thus one can anticipate continued efforts to establish effective long-term remedial programs.

References: Chapter 4

BPA (Bonneville Power Administration), 1984, *The Expanded Residential Weatherization Program, Final Environmental Impact Statement,* DOE/EIS-0095F, Vol. 1, p. A.15.

Chruscielewski W., Domanski T. and Orzechowski W., 1983, "Concentrations of radon and its progeny in the rooms of Polish spas," *Health Phys.* **45**, 421.

Cohen B. L., 1982, "Health effects of radon emissions from uranium mill tailings," *Health Phys.* **42**, 695.

Crawford M., 1985, "Mill tailings: A $4-billion problem," *Science* **229**, 537.

Chemical Rubber Publishing Company, 1978, *Handbook of Chemistry and Physics.* 58th Edition 1977-78 (Cleveland: Chemical Rubber Publishing Co).

EPA (Environmental Protection Agency), 1983, "Environmental Standards for Uranium and Thorium Mill Tailings at Licensed Commercial Processing Sites; Final Rule," 40 CFR Part 192, *Federal Register* **48**, 45926 (October 7).

Gesell T. F., 1983, "Background atmospheric ^{222}Rn concentrations outdoors and indoors: a review," *Health Phys.* **45**, 289.

Guimond R. J., Jr., Ellett W. H., Fitzgerald J. E., Jr., Windham S. T. and Cuny P. A., 1979, *Indoor Radiation Exposure Due to Radium-226 In Florida Phosphate Lands,* U.S. Environmental

Protection Agency, Office of Radiation Programs, EPA 520/4-78-013 (Washington, DC: EPA).

Hartley J. A., Koehmstedt P. L., Esterl D. J. and Freeman H. D., 1980, "Uranium mill tailings stabilization," in: *Waste Management 80* (edited by Roy G. Post), pp. 193-204 (Tucson: Univ. of Arizona).

Hess C. T., Weiffenbach C. V., Norton S. A., Brutsaert W. F. and Hess A. L., 1982, "Radon-222 in potable water supplies in Maine: The geology, hydrology, physics, and health effects," in: *Natural Radiation Environment* (edited by K. G. Vohra *et al.*), pp. 216-220 (New Delhi: Wiley Eastern).

Joseph A. B., Gustafson P. F., Russell I. R., Schuert E. A., Volchock H. L. and Tamplin A., 1971, "Sources of Radioactivity and Their Characterization,' in *Radioactivity in the Marine Environment*, National Academy of Sciences (Washington, DC: NAS), pp. 6-41.

NCRP (National Council for Radiation Protection and Measurements), 1975, "Natural Background Radiation in the United States," *NCRP Report No. 45* (Bethesda, MD: NCRP).

_____, 1976 "Environmental Radiation Measurements," *NCRP Report No. 50* (Bethesda, MD: NCRP).

_____, 1984a, "Exposures from the Uranium Series With Emphasis on Radon and its Daughters," *NCRP Report No. 77* (Bethesda, MD: NCRP).

_____, 1984b, "Evaluation of Occupational and Environmental Exposures to Radon and Radon Daughters in the United States," *NCRP Report No. 78* (Bethesda, MD: NCRP).

Pohl R., 1975, "Nuclear energy: Health efects of Thorium-230," unpublished manuscript.

Pohl-Ruhling J., Steinhauser F. and Pohl E., 1982. "Radiation exposure and resulting risk due to residence and employment in radon spa," in: *Natural Radiation Environment*, (edited by K.G. Vohra et. al.), pp. 107-113 (Wiley Eastern).

Pohl R., 1982, "Will It Stay Put?" *Physics Today* **35**, No. 12, p. 37.

Rogers V. C. and Nielson K. K., 1980, "Tailings piles from uranium producers," in: *Waste Management 80*, (edited by Roy G. Post), pp. 163-172 (Tucson: Univ. of Arizona).

Roessler C. E., Roessler G. S. and Bolch W. E., 1983, "Indoor radon progeny exposure in the the Florida phosphate mining region: a review," *Health Phys.* **45**, 389.

Chapter 5

INDOOR RADON LEVELS

Maurice A. Robkin

A. Introduction

Uranium is one of the most widely spread of all of the elements. It occurs, with considerable variation in concentration, at an overall low average value. It is present in all rocks and soil and therefore in most of the raw materials from which we process finished products. As a result, its daughter, radium, is also widely spread, particularly in those products which are made from mineral products. Radium, with a half-life of 1600 years, continuously generates its immediate daughter, radon.

Radon, being a noble gas with a half-life of 3.8 days, can diffuse for some distance through porous materials before decaying to its short lived daughters. When it is generated in or near to buildings, it can diffuse into living spaces. The concentration of radon and its daughters in living spaces depends on the balance between the rate at which they are removed from the air and the rate at which they are introduced. The factors that determine these rates vary greatly from one location and building to another, depending on the site and on details of the building construction.

The main source of radon in buildings is the ground underneath the building. Other sources include building materials, domestic water, and gas supplies. These individual contributions are discussed in the following Section.

B. Sources of indoor radon

1. Building materials. Building materials that are made from stone, sand, ore byproducts and the like, contain uranium and radium and generate radon. Many of these materials such as brick, wallboard or concrete are sufficiently porous to allow the radon to escape into the air. Table 5-1 summarizes, from studies carried out in the United Kingdom, Russia, West Germany, Spain and the United States, the radium-226 concentrations found in various building materials. It is seen that for all of the indicated materials the radium-226 content varies widely, except for granite which is relatively homogeneous and has a high radium-226 content. Materials which do not derive from the earth's crust, such as wood, tend to have a very low radium concentration.

This set of data is not inclusive, and it may be expected that actual variations are even greater than indicated here. For example, in the case of natural gypsum, Table 5-1 shows that its radium-226 concentration is lower than that of many other materials, at quoted levels of 0.6 pCi/g or less. However, phosphogypsum, a by-product of phosphate production, can have concentrations of over 30 pCi/g (Colle 1981).

TABLE 5-1. *Radium-226 concentrations in selected building materials, from studies in several countries (in pCi per gram).*

Material	United Kingdom (Ham71)	USSR (Kr71)	West Germany (Ko74)	Spain (Ga82)	USA (Kah83)	USA* (In83)
Gypsum	0.6		0.5	0.08		0.34
Sand, Gravel		0.4-1	0.4	0.38-0.62		
Bricks	0.2-1.4	0.5-1.5	1.7	0.92	0.2-3.5	0.8-1.2
Concrete		2.0	0.6		0.7-1.4	
Cement		0.7	0.5	2.04		
Tile				1.9	1.7-1.9	
Granite	2.4	3.0	2.8	2.09		

*Quoted concentrations are for uranium-238 (not radium-226).

The rate at which radon emanates from various materials is indicated in Table 5-2. The rates are normalized to the radium-226 concentration. For a given radium-226 concentration, the radon emanation rate depends on the physical properties of the material. For example, soil, being highly porous, has an emanation rate 100 times greater than concrete. Emanation rates are also affected by various meteorological conditions. The most interesting are the effects of moisture content and barometric pressure.

TABLE 5-2. *Relative radon emanation rates from various materials.*

Material	Thickness (cm)	Relative Rate*
Concrete	10	0.005
Light concrete	20	0.02
Heavy concrete	8	0.01
Phosphogypsum	1.3	0.001
Phosphogypsum	7.6	0.01
Soil	infinite	0.5
Uranium mill tailings	10	0.2
Uranium mill tailings	infinite	1.6

Reference: Colle 1981.
*Ratio of radon-222 emanation rate (pCi per square-meter per second) to radium-226 concentration (pCi per gram).

Increase in pore moisture content increases the emanation rate, at least at low moisture levels. In one hypothesis, this has been explained

(Colle 1981) as due to the increase of the energy loss, in the water in the pores, of radon recoiling from the decay of radium in the bulk material. This increases the amount of radon deposited in the pores, which enhances diffusion. In another hypothesis, the suggestion has been made (Ingersoll 1983) that water in the pores leaches radon from a crystalline structure that is damaged by alpha decay. In either event, an increase in the moisture content of a few percent can result in a six-fold increase in the emanation rate. It has been pointed out, however, that the data are very selective and based on a very limited set of results (Colle 1981).

The increase in emanation rate with moisture content may occur only when the water content of the pores is low. Other workers (Tanner 1964; Cox 1970) indicate that the emanation rate of radon from soil decreases as the water content of the soil increases. This happens when the water filling the pores interferes with the diffusion of radon through the pores. The indications are that the crossover point between enhancement and inhibition occurs at a relatively low level of pore water content.

Ingersoll has measured the emanation rates from various samples of building materials, using disk-shaped test pieces (Ingersoll 1983). The test pieces were 6 inches in diameter and one inch thick. These pieces are too small to yield the asymptotic emanation rate from an infinite slab of material, but they do give a comparison of the relative emanation rate from the different materials. In particular, the average escape to production ratio allows an ordering of the different materials in terms of their potential rate of emanation of radon. Normalized to the same radon production rate in the material, gypsum and soil samples gave about the same emanation rate, but the emanation was only 1/25 as much for a red brick sample.

Pressure effects are also important. A 1% to 2% decrease in air pressure can double the emanation rate (Colle 1981). This is a "pumping" effect, which is also seen for emanation from soil.

2. Ground water. Radon and radium are soluble in water. When ground water moves through radium-bearing soil and rocks, they are dissolved and transported with the water. As radon is transported, it and its daughters decay, so that for radon only the most recent history of the movement of the water is significant. If there is enough radium dissolved in the water to be the dominant contributor to the local radon concentration, then the entire history of the water movement is important.

When radon-containing ground water is used for domestic purposes, the radon it contains can be released into the living space. The amount of radon available for release depends on the type and degree of processing which the water undergoes before domestic use.

In a survey taken over the United States (Hess 1985), radon levels in water were found to vary over a large range, with the highest values in Northeast and Southwest states and the lowest generally in South-

Central and Southeast states. Often private wells have considerably higher radon concentrations than public supplies, but this is not a consistent result. Even for a given type of rock, there can be large variations about the average. For example, Hess *et al.* (1985) found radon concentrations of up to 300,000 pCi/ℓ in granite, with an average value of 22,000 pCi/ℓ. There are similarly large variations for the radon in water supplies. In measurements of radon concentrations in 6310 public ground water supplies, about 75% had levels below 500 pCi/ℓ, but about 2% exceeded 10,000 pCi/ℓ, and 0.5% exceeded 100,000 pCi/ℓ (Hess 1985).

The adverse health effects of radon in water are largely due to the transfer of radon to the air, where it decays and its short lived radioactive daughters inhaled, rather than from direct ingestion (Cross 1985). Thus it is important to know the contribution which radon in the water supply of a given house makes to the average radon concentration in the air.

Gesell and Prichard (1980) performed measurements in an apartment when it was vacant and immediately after the occupants returned and began to use water. A very clear correlation was seen between water usage and indoor radon level, with the radon concentration in the air rising from a plateau of about 0.5 pCi/ℓ in the absence of water use to peaks in the neighborhood of 2 pCi/ℓ when water was used. The water supply for this apartment had a relatively high radon concentration, 1500-2000 pCi of radon per liter of water. An apartment on an upper floor is isolated from radon inflow from the ground and is more sensitive to internal sources.

Recent studies have adopted an average value of about 0.1 pCi/ℓ of radon in air per 1000 pCi/ℓ of radon in water (Nazaroff 1987; Hess 1985; Kahlos 1980). Differences in patterns of water use, in ventilation, and in house arrangement will cause wide departures from this average. The typical level for water supplies in the United States is of the order of 1000 pCi/ℓ or less, indicating that on average the water supply contributes less than 0.1 pCi/ℓ to the radon indoor air concentration. Thus, water supplies are usually not a major factor in determining indoor air concentrations of radon although there can be appreciable contributions in some cases, especially when water is drawn from private wells. For this latter group, the average contribution from water is estimated to be 0.6 pCi/ℓ; it should be noted that the average value is so high not because 0.6 pCi/ℓ is a "typical" value but because a small fraction of these homes have much higher levels.

3. Natural gas. Natural gas, since it is produced from underground reservoirs, also contains radon. When it is burned in domestic appliances, radon is released into the living space. The amount of radon in natural gas at the point of consumption can be considerably less than the amount present at the well head due to the decay en route or while the gas is held up in storage.

In the case of natural gas, radon concentrations ranging from 0.8

pCi/ℓ at the onshore distribution manifold for North Sea gas (Wilkins 1980) to 1450 pCi/ℓ for North American gas at the wellhead (Gesell 1975) have been reported. There can be very wide variations even for gas from the same area. Thus, in the Texas Panhandle, levels of under 5 pCi/ℓ have been found as well as the 1450 pCi/ℓ value cited above.

If it is assumed that gas space-heating and water-heating units are vented to the outside, unvented cooking ranges provide the main channel by which radon from natural gas will enter the home. The average gas range uses roughly 8000 cubic feet of gas a year, or, on average, about 1 cubic foot per hour (AGA 1981). The concentration of radon would be expected to be greatest in the kitchen but, for purpose of an order-of-magnitude estimate, it will be assumed to be distributed over a living area with 10,000 cubic feet of air. If this space is ventilated at the rate of 1 air change per hour, then the cubic foot of natural gas used per hour will be diluted by a factor of 10,000. In consequence, for every 10,000 pCi/ℓ radon in the input gas, there will be an average concentration of 1 pCi/ℓ of radon in the indoor air. Thus, even for the case of 1450 pCi/ℓ of radon in natural gas, cited above as an unusually high concentration, the resulting average indoor radon concentration for a house with an unvented gas stove will be only 0.1 to 0.2 pCi/ℓ. Averaged over the full housing stock, and considering the decay of radon during its transport in the pipeline to the consumer, the contribution is much lower.

Thus, natural gas is not a significant contributor to indoor radon on an overall basis, although it might be in isolated cases, for example, in a kitchen which has little air exchange with either the outdoors or the remainder of the house.

4. Soil. Under normal circumstances the dominant contributor to indoor radon concentrations is the emanation from soil (Nero 1985; Hess 1985). The emanation rate varies with the local concentration of radium and with the atmospheric pressure. When the pressure rises, air with low concentration of radon is forced into the soil. When it falls, the air in the soil containing high radon levels is "exhaled" into the air. Thus, the maximum rate of soil emanation is correlated with low pressure. In addition, a house sitting on the ground acts like a chimney, drawing soil gas into itself from beyond its own perimeter as well as from immediately beneath.

The emanation from the soil allows radon to diffuse into buildings directly from the ground on which the building sits. Many houses are built with a concrete foundation with the enclosed soil covered with a thick concrete pad. The emanation rate from concrete is much less than from the soil and the diffusion of radon through concrete is very small. However, unless there is very careful sealing of the building, radon can find portals of entry through cracks in the foundations, along pipe penetrations, from floor drains, etc. A Bonneville Power Administration study finds emanation rates from bare soil are reduced by about a factor of ten if the soil is covered with concrete (BPA 1984).

5. Summary. When the problem of indoor radon first began to attract detailed attention, sources other than the underlying ground at first were thought to be quite important. However, subsequent detailed studies have shown that, for private houses, radon from the soil accounts for most of the indoor concentration. In fact, considering overall averages, Nero (1985) attributes virtually all of the excess over the outdoor level to the soil contribution. There are exceptions for specific houses and the situation is quite different for apartments, which in general are more isolated from the ground and have lower radon concentrations. Identification of the ground as the dominant source has important implications for the sorts of remedial actions which should be taken (see Chapter 6).

C. Observed typical levels

1. Measured levels. In this Section we consider indoor radon concentrations which are typically found under usual natural conditions, that is, in regions where there have been no human activities or special natural conditions (such as radioactive deposits) which lead to enhanced levels. In the next Section, a few regions are discussed where high radon levels have been observed.

In order to assess the importance of the indoor radon problem on a national scale, it is convenient to refer to a nominal average value of the indoor radon concentration. Such averages have been used by a number of authors. For example, Report No. 78 of the National Council on Radiation Protection and Measurements (NCRP 1984) adopted George and Breslin's (1980) value of 0.83 pCi/ℓ obtained from measurements on 21 houses in the New York/New Jersey area. Reiland *et al.* (1985) quote a mean level during the heating season of 1.3 pCi/ℓ for a sample of 146 homes built to current building codes in the Pacific Northwest. During the winter heating season, householders normally reduce the infiltration rate of outside air so that, during this period, houses would have less than their annual average ventilation rate.

One problem in determining a representative value for the radon concentration is the method by which the data were obtained. In order to fairly represent the housing stock, there should not be any bias in the selection of the houses in which to measure the radon concentration. There have been two recent studies of single-family houses in which the selection of the houses was uncorrelated with their radon concentrations. That is, the sample was unbiased. A Lawrence Berkeley Laboratory study of the literature on radon measurements (Nero 1984) found a geometric mean value of 0.96 pCi/ℓ and an arithmetic mean value of 1.66 pCi/ℓ for a sample of 552 houses. These were not preselected on the basis of a suspicion of elevated radon levels. In a study of 453 homes of physics professors (a sample picked for convenience in data collection), Cohen (1986) found a geometric mean concentration of 1.0 pCi/ℓ and an arithmetic mean of 1.47 pCi/ℓ. Together these studies point to approximate averages of 1.0 pCi/ℓ (37 Bq/m^3) for the geometric mean and 1.5 pCi/ℓ (56 Bq/m^3) for the arithmetic mean for single-family homes.

The results of these and earlier studies show a characteristic distribution for the number of houses as a function of radon concentrations, the so-called lognormal distribution (George 1983; Ryan 1983; Alter 1983; Nero 1984; Cohen 1986). In such a distribution there is a long "tail" in which an appreciable number of houses have concentrations far above the average, as illustrated in Figure 5-1. As can be seen, the range is considerable, extending from under 0.2 pCi/ℓ to over 8 pCi/ℓ.

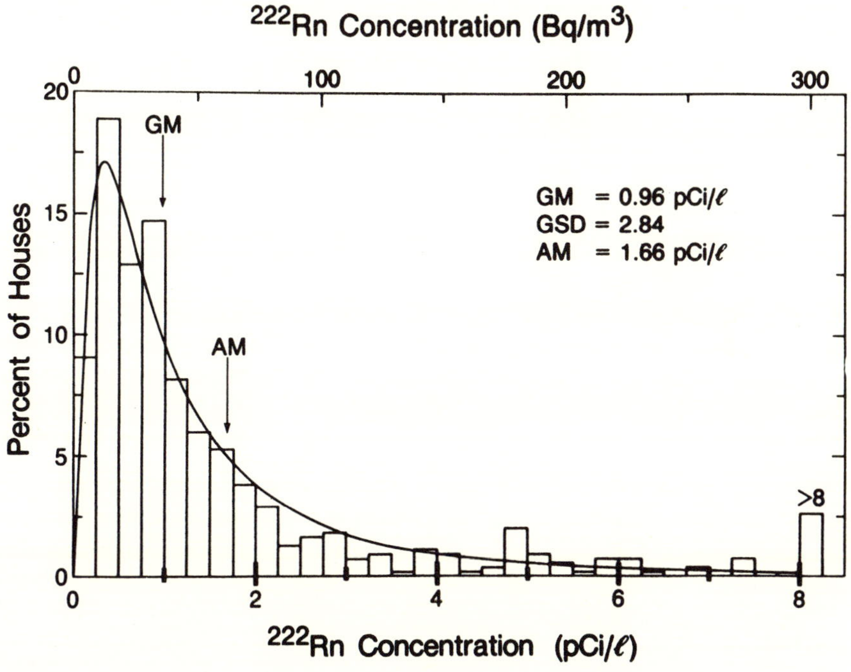

FIG. 5-1. Percent of single-family homes with different values of radon-226 concentration. The bar graph gives observed numbers: the solid curve is a graph of a lognormal distribution with a geometric mean of 0.96 pCi/ℓ and a geometric standard deviation of 2.84 (where the parameters are chosen to fit the observed data). The arithmetic mean for this distribution is 1.66 pCi/ℓ. The sample includes 552 homes from 19 studies. It excludes studies in regions where unusually high radon concentrations had been suspected. [Reprinted, with permission, from Nero et al., LBL-18274 (Nero 1984, Figure 1).].

The sample sizes for both the Nero and the Cohen surveys were small. A much larger data set has been reported by Alter and Oswald (Alter 1987). Unfortunately, the data in their set were self-selected. That is, they report a summary of the measurements obtained from householders who elected to use the services of a commercial radon measurement company. While their data base is very large, containing over 60,000 measurements, there were a very large number from the states of Pennsylvania (22312), Washington (11711), Oregon (5835), New York (4906) and New Jersey (4475).

The State of Pennsylvania is associated with the Reading Prong (see Section D), and one might reasonably expect that the publicity associated with this discovery would lead local householders to want to have their houses measured for radon. Thus, the contribution to the data base from Pennsylvania may bias it toward higher values. The Reading Prong is known to extend through New Jersey and New York. The same effect from publicity may upwardly bias the data from those states.

Conversely, the large number of samples from Washington and Oregon is due to a survey carried out by the Bonneville Power Administration in connection with its residential energy conservation program (BPA 1984). Western Washington and Oregon have lower average radon levels than most of the country, biasing the sample in a direction opposite to that of the Reading Prong. In general, however, one would expect that the main effect would be a bias to higher values. That is, once a high reading is made public there would be a tendency for local public concern to stimulate more testing.

The average radon concentration from these measurements, with no corrections for bias, is over 7 pCi/ℓ. The authors have re-analyzed the data using a number of approaches designed to reduce the bias and each of these approaches gives an arithmetic average concentration of about 4 pCi/ℓ. It appears unlikely that the bias has been fully eliminated and that this latter value is definitive. However, the results highlight the need for a comprehensive program of unbiased (not preselected) measurements around the country.

Hess *et al.* (1985) summarized a number of surveys of the indoor radon levels in "background" houses not associated with elevated soil burdens of radium. Their review spans a large number of dwellings across the country and their data are shown in Table 5-3. It is to be noted that the levels are highest in basements and lowest on the second floor and that they are elevated in bathrooms (presumably due to water usage). In the climate of Pennsylvania the levels are higher in winter than summer (presumably due to ventilation differences).

Nero *et al.* (1983) have observed the effect of ventilation on the indoor radon level in a group of houses in the San Francisco area. Their results show a clear downward trend of radon concentration with increasing ventilation rate. However, from their data, they conclude that most of the variance in the observed concentrations is due to the source term; i.e., from the variation in the leakage rate of radon from

TABLE 5-3. *Comparisons of indoor radon concentrations in different parts of houses (geometric means).*

Characterization of houses studied		Season	Radon Concentration (pCi/ℓ) Bsmnt	1st Fl	2nd Fl	Bathrm
New York-New Jersey	(1)		1.7	0.83	0.7	
Central Maine	(2)		2.46	1.40	1.12	1.62
Houston	(3)			0.39		0.58
Eastern Pennsylvania	(4)	Summer	3.40	1.22		
		Winter	5.90	4.40		
Northeastern U.S.	(5)					
Non-efficient		Full year	1.3	0.3		
Energy efficient		Full year	3.57	1.60	2.01	

Notes:

1. From literature survey by Hess *et al.* (1985): 18 ordinary houses; basements studied only for 9.
2. From literature survey by Hess *et al.* (1985): 67-82 ordinary houses, depending upon area studied.
3. From literature survey by Hess *et al.* (1985): 81-103 houses and apartments.
4. From literature survey by Hess *et al.* (1985): 36 houses.
5. From Fleischer *et al.* (1983), Table 5: 3 non-energy efficient houses, 7 energy efficient houses; (results converted to geometric means).

the soil into the houses.

The conclusion which can be drawn is that, for a given source rate and for low ambient outside air concentrations of radon, the radon concentration in houses is inversely proportional to the ventilation rate but that the primary determinator of the levels in houses is the source term. As corollaries, since the primary source is from radon in soil gas, the average concentration in a basement is greater than elsewhere in a house and the average concentration in multi-story apartment houses is lower than in single-family houses (Nero 1985).

2. Specification of average radon concentrations. There are many difficulties with specifying and using an "average" level. These difficulties include: 1) the existing surveys of radon concentrations around the country are limited in extent; 2) it is not certain that the samples studied are truly representative; 3) the large variations observed within a given house suggests that there should be a weighting of the concentration in a given part of the house by the product of number of occupants times exposure time; and 4) the variations over time, throughout the year, should be accounted for.

Estimation of the average itself requires caution. As discussed in the preceeding section, radon concentrations in dwellings are usually rather well described by a lognormal distribution. Commonly cited averages include the arithmetic mean, the geometric mean, and the median. For a lognormal distribution, the geometric mean is equal to the median and the arithmetic mean is greater than either. Lognormal distributions are characterized by a long tail at high values, corresponding to occasional very high radon concentrations.

Which mean is the most appropriate depends on what use is to be made of it. For example, if one assumes that health effects are linearly related to dose (see Chapter 9), then the arithmetic mean is relevant with the proviso that some sort of population weighting may be needed. The median, the concentration value which divides the lower half of the values from the upper half, and the mode, the concentration value at the peak of the distribution, serve to characterize the distribution and provide some means of distinguishing the data from one locale from that of another, as well as in some sense indicating "typical" values. The geometric mean and its variance, if one is dealing with a lognormal distribution, specify the shape of the distribution curve and of the occurrence of concentrations much greater than the arithmetic average. This latter point is important because even a small percentage of elevated values may still impact a large number of people.

As stated above, the geometric mean of the radon concentrations in typical single-family houses across the United States is about 1.0 pCi/ℓ and the arithmetic mean is about 1.5 pCi/ℓ. It is not far off, according to this present information, to use 1 pCi/ℓ as a rough "average" value for all U.S. houses, including apartments.

3. Equilibrium factors. The significant quantity insofar as health effects are concerned, is not the radon concentration per se, but the concentration of its short lived daughters, usually expressed in working levels (WL), and their distribution between the attached and the unattached fractions. Because of the enhanced deposition in the lung of the unattached fraction, its contribution to the lung dose, per unit contribution to the working level, is considerably greater than that of the attached fraction. However, no distinction is made between the two fractions in determining the working level.

The number of working levels in air for a given concentration of radon depends on a number of local factors and is usually expressed in terms of the "equilibrium factor", F (See Chapter 2, Section G). As will be discussed below, an equilibrium factor of about 0.5 seems to represent a great deal of the data from various reports and has some theoretical support (Swedjemark 1983), although in a recent report of the National Council on Radiation Protection and Measurements a value of 0.7 was used (NCRP 1984b). If measurements are made which directly establish the radon daughter WL (rather than the radon concentration) knowledge of the equilibrium factor is unnecessary, but this is not a common practice.

D. Observed elevated levels

The dominant source for indoor radon levels is the rate of emanation of radon from the soil beneath the house. This rate varies with the concentration of radium in the soil which in some locations can be quite high. High emanation rates are correlated with human activities which result in local concentrations of radium and with high local concentrations of uranium in permeable ground. The activity of radium in the ground will be in equilibrium with the activity of uranium unless there has been differential solution and removal over time of one species relative to the other.

Very high levels of indoor radon are being found in some regions of the United States. Prominent among these is a region of Pennsylvania called the "Reading Prong". This geological feature extends in a Northeasterly direction from the vicinity of Reading, Pennsylvania through Easton, New Jersey and into New York. The discovery of these high radon levels was made when an engineer at the Limerick Nuclear Power Plant tripped radiation alarms when he reported to work. The contamination was traced to high levels of radioactivity in his home.

Subsequent studies of 18,000 homes in the area by the Pennsylvania Department of Environmental Resources have shown that 59% of the homes have measured levels above 0.02 WL, corresponding to a concentration of 4 pCi/ℓ for F = 0.5. This is the level at which the Environmental Protection Agency recommends that corrective action be taken (EPA 1986). Further, 12% have levels above 0.1 WL, 0.6% above 1 WL, and at least one above 10 WL (Gerusky 1987).

These high values are not confined to Pennsylvania, as indicated by the national data of Alter and Oswald (1987). Over 29% of the surveyed homes had concentrations above the EPA remedial action level of 4 pCi/ℓ. The highest single radon concentration in their data set was a value from Maine of 4354 pCi/ℓ. For an equilibrium value of 0.5, this corresponds to about 22 WL. A level of 20 WL gives an annual exposure which is 200 times the acceptable value for occupational exposure of uranium miners, although it is to be noted that such an estimate of the annual exposure could be misleading unless the measurements reflect the seasonal variations over a full year. These high levels are explained in terms of uranium deposits which come close to the surface.

Uranium in concentrations elevated from the average are also found in association with phosphate deposits. Where the ores extend near to the surface or they or their waste products are deposited on the surface, the associated radium-generated radon can diffuse into the air. Production of phosphate fertilizers results in the generation of gypsum wastes into which the uranium and its daughters have been segregated. Florida produces phosphate and structures are built on reclaimed lands in which there are elevated radium levels and elevated emanation of radon.

Roessler et al (1983) have reviewed the levels of radon in structures associated with the phosphate lands in Polk County, Florida. The levels of radon daughters in structures correlates with the radon levels in the various categories of soil. About 600 structures in Polk County, less than 1% of the County wide total, were found to have exposure levels in excess of 0.02 WL. In a study specifically of houses built on reclaimed land, 7 of 133 houses had levels above 0.05 WL (Guimond 1979).

George and Eng (1983) studied 33 structures in New Jersey, New York and Pennsylvania that were believed to be uncontaminated but were located near to former industrial sites at which elevated levels of radium in the soil were found. The levels of radon daughters in these structures were compared to values in structures located far away from these sites. Their results showed no strong differences and, in fact, the values for the structures near the New York and New Jersey sites are somewhat lower than the others. The geometric average level in the 33 houses studied was about 0.003 WL with a range from 0.002 to 0.027 WL.

Hawthorne *et al.* (1985) measured the indoor air quality in 40 homes in East Tennessee. They observed a difference in the radon concentration in these homes according to their location. A home located in a valley tended to have a lower concentration than a home on a ridge. This was ascribed to the presence of a deeper overburden of low activity soil on the valley floor. The ridges are of dolomite with relatively shallow soil cover which was thought to offer relatively lower resistance to diffusion of radon from the underlying uranium bearing shale. For two houses on the ridge, the radon concentrations were as high as 13 pCi/ℓ. Cohen suggests (1986b) that a better explanation is that houses on hills are subjected to higher wind velocities and therefore to greater induced pressure drops inside of the house (a phenomenon called the "Bernoulli effect"). This reduced pressure enhances the inflow of radon gas from the soil underlying a hilltop house.

Fleischer *et al.* (1983) have explored the effect on the radon concentration of making a home energy efficient. Their results are incorporated in Table 5-3. The authors concluded that "energy-efficient homes tend to be significantly more radon-rich than are the more conventional homes," especially in the living areas. This was attributed to the houses being tighter and, in some cases, to the use of solar heating with radon-emitting rock or sand for heat storage. In their study, they found one house with a basement level of 200 pCi/ℓ of radon. In this case the elevated level was traced to the venting of air from a well into the basement. The high level was eliminated by eliminating the well venting.

A well known case in which human activities resulted in elevated indoor levels of radon and its daughters involved the use of uranium mill tailings as construction materials and land fill in Western Colorado in the 1950s and 1960s. Studies carried out in Grand Junction,

Colorado (Siek 1972) indicated that in 398 indoor locations sampled for radon daughters, 44% were above 0.01 WL and 19% were above 0.05 WL. The normal background in the area in places not associated with local concentrations of tailings was measured to be about 0.004 WL.

E. Summary

The main source of indoor radon is direct emanation of radon gas from the soil into houses. This inflow is enhanced by the pressure balance of the house itself which actually draws soil gas into the air of the lowest floor which is often a basement. From the basement, the area of highest concentration, it is distributed throughout the remainder of the house.

In special cases, such as the upper floors of apartment buildings in localities where the concentration of radium in water is high, domestic water usage can provide a significant source of radon. However, for single-family homes the direct emanation from the soil is the most important source.

Survey data of many dwellings indicate that the concentration of radon tends to follow a log-normal distribution with a geometric average concentration for single-family houses of about 1 pCi/ℓ and an arithmetic average concentration of about 1.5 pCi/ℓ. A typical equilibrium factor is about 0.5 so that the arithmetic average corresponds to about 0.0075 WL. However, many homes have much higher levels, some as high or higher than 1 WL which can produce lung exposures many times higher than what is permissable for uranium miners and many times higher than the corrective action level of 0.02 WL recommended by the EPA. Clearly further surveys are needed, both to identify such houses and to investigate the suggestion (Alter 1987) that the average concentrations considerably exceed those cited above.

References: Chapter 5

AGA (American Gas Association), 1981, *Gas Facts, 1980 Data,* p. 135 (Arlington: AGA).

Alter H. W. and Oswald R. A., 1983, "Results of indoor radon measurements using the track etch (R) method," *Health Phys.* **45**, 425.

______, 1987, "Nationwide distribution of indoor radon measurements: a preliminary data base," *J. Air Pollution Contr. Assoc.* **37**, 227.

BPA (Bonneville Power Administration), 1984, *The Expanded Residential Weatherization Program, Final Environmental Impact Statement*, DOE/EIS-0095F, Vol. 1, p. A. 15.

Cohen B. L., 1986a, "A national survey of ^{222}Rn in U.S. homes and correlating factors," *Health Phys.* **51,** 175.

______, 1986b, private communication.

Colle R., Rubin R. J., Knab L. I. and Hutchinson J. M. R., 1981, "Radon Transport Through and Exhalation from Building Materials: A Review and Assessment," NBS Tech. Note 1139.

Cox W. M., Blanchard R. L. and Kahn B., 1970, "Relation of Radon Concentration in the Atmosphere to Total Moisture Detention in Soil and Atmospheric Thermal Stability," in: *Radionuclides in the Environment,* ACS Adv. Chem. Series No. **98**, American Chemical Society, 436-446.

Cross F. T., Harley N. F. and Hofmann W., 1985, "Health effects and risks from ^{222}Rn in drinking water," *Health Phys.* **48,** 649.

EPA (Environmental Protection Agency) and Department of Health and Human Services, 1986, *A Citizen's Guide to Radon, What It Is and What To Do About It*, OPA-86-004 (August 1986).

Fleischer R. L., Mogro-Campero A. and Turner L. G., 1983, "Indoor radon levels in the Northeastern U.S.: Effects of energy-efficiency in homes," *Health Phys.* **45**, 407.

Garzon L., Fontenla P. and Suarez A., 1982, "Radioactivity of Building Material-Absorbed Doses," in *Natural Radiation Environment* (edited by K. G. Vohra et al.) (Wiley Eastern Ltd).

George A. C. and Eng J., 1983, "Indoor radon measurements in New Jersey, New York and Pennsylvania," *Health Phys.* **45**, 397.

George A. C. and Breslin A. J., 1980, "The Distribution of Ambient Radon Daughters in Residential Buildings in the New Jersey-New York Area" in *Natural Radiation Environment III, Vol 2*, CONF-780422, 1272 (Oak Ridge Technical Information Center/U.S. D.O.E.)

Gerusky T. M., 1987, "Pennsylvania: protecting the homefront," *Environment* **29**, No. 1, 12.

Gesell T. F., 1975, "Occupational radiation exposure due to Rn-222 in natural gas and natural gas products," *Health Phys.* **29**, 681.

Gesell T. F. and Prichard H. M., 1980, "The Contribution of Radon in Tap Water to Indoor Radon Concentrations" in *Natural Radiation Environment III, Vol 2*, CONF-780422, 1347-1363 (Oak Ridge Technical Information Center/U.S. D.O.E.).

Guimond, R. J., Jr., Ellett W. H., Fitzgerald, J. E., Jr., Windham S. T. and Cuny P. A., 1979, "Indoor Radiation Exposure Due to Radium-226 In Florida Phosphate Lands," EPA 520/4-78-013, U.S.E.P.A. Office of Radiation Programs.

Hamilton E. I., 1971, "Relative radioactivity in building materials," *Am. Ind. Hyg. Assn. J.* **32**, 398.

Hawthorne A. R., Gammage R. B. and Dudney C. S., 1985, "An indoor air quality study of forty East Tennessee homes," paper submitted to: *Environment International,* private communication to M. Robkin.

Hess C. T., Weiffenbach C. V. and Norton S. A., 1983, "Environmental radon and cancer correlations in Maine," *Health Phys.* **45**, 339.

Hess C. T., Michel J., Horton T. R., Prichard H. M. and Coniglio

W. A., 1985, "The occurrence of radioactivity in public water supplies in the United States," *Health Phys.* **48**, 553.

Ingersoll J. G., 1983, "A survey of radionuclide contents and radon emanation rates in building materials used in the U.S.," *Health Phys.* **45**, 363.

Kahlos H. and Asikainen M., 1980, "Internal radiation doses from radioactivity of drinking water in Finland," *Health Phys.* **39,** 108.

Kahn B., Eichholz G. G. and Clarke F. J., 1983, "Search for building materials as sources of elevated radiation dose," *Health Phys.* **45**, 349.

Kolb W., 1974 "Influence of building materials on the radiation dose to the population," *Kernenergie und Offenlichkeit* 4, 18.

Krisiuk E. M., Karasov S. I., Shamov V. P., Shalak N. I., Lisachenko E. P. and Gomelsky L. G., 1971, *A Study of Radioactivity in Building Materials,* Research Institute for Radiation Hygiene, Leningrad.

NCRP (National Council on Radiation Protection and Measurements), 1984, "Evaluation of Occupational and Environmental Exposures to Radon and Radon Daughters in the United States," *NCRP Report No. 78* (Bethesda, MD: NCRP).

Nazaroff W. W., Doyle S. M., Nero A. V. and Sextro R. G., 1987, "Potable water as a source of airborne ^{222}Rn in U.S. dwellings: a review and assessment," *Health Phys.* **52**, 281.

Nero A. V., Boegel M. L., Hollowell C. D., Ingersoll J. G. and Nazaroff W. W., 1983, "Radon concentrations and infiltration rates measured in conventional and energy-efficient houses," *Health Phys.* **45**, 401.

Nero A. V., Schwehr M. B., Nazaroff W. W. and Revzan K. L., 1984, *Distribution of airborne ^{222}Radon concentrations in U.S. homes,* Lawrence Berkeley Laboratory, Berkeley, CA 94720, LBL-18274.

Nero A. V., Jr., 1985, *Indoor Concentrations of Radon 222 and Its Daughters: Sources, Range, and Environmental Influences,* Lawrence Berkeley Laboratory, Berkeley, CA 94720, LBL-19346.

Reiland P., McKinstry M. and Thor P., 1985, "Preliminary Radon Testing Results for the Residential Standards Demonstration Program," Bonneville Power Administration (unpublished).

Roessler C. E., Roessler G. S. and Bolch W. E., 1983, "Indoor radon progeny exposure in the Florida phosphate mining region: a review," *Health Phys.* **45**, 389.

Ryan M. T., Goldsmith W. A., Poston J. W., Haywood F. F. and Witherspoon J. P., 1983, *Radon Dosimetry: A Review of Radon and Radon Daughter Exposure Conditions in Dwellings and Other Structures,* Oak Ridge National Laboratory, Oak Ridge, TN 37830, ORNL/TM-5286.

Siek R. D., 1972, Testimony reported in: *Use of Uranium Mill Tailings for Construction Purposes,* Hearings Before the Subcommittee on Raw Materials of the Joint Committee on Atomic Energy, 92nd

Congress, 1st Session, pg. 179-233.

Swedjemark G. A., 1983, "The equilibrium factor F," *Health Phys.* **45**, 453.

Tanner A. B., 1964, "Radon Migration in the Ground: A Review," in *The Natural Radiation Environment* (edited by J. A. S. Adams and W. H. Lowder), pp. 101-190, (Chicago: Univ. of Chicago Press).

Wilkins B. T., 1980, "The Assessment of Radon and its Daughters in North Sea Gas Used in the United Kingdom." in: *Radiation Protection: A Systematic Approach to Safety*, Vol. 2, Proceedings of the 5th Congress of the International Radiation Protection Society, Jerusalem, March, 1980 (Oxford: Pergammon), pp. 1143-1148.

Chapter 6

MODIFICATION OF RADON LEVELS IN HOMES

Peter A. Breysse

A. Radon entry mechanisms

It is important that sources of radon as well as radon infiltration mechanisms be understood before attempts are made to control indoor radon levels. The overall situation has been summarized as follows (Nazaroff 1984):

> The radon concentration indoors is determined by a balance between the rate of entry from sources and the rate of removal, primarily by ventilation. The results of measurements in several countries show that the ventilation rate is more narrowly distributed than either the radon concentration or the radon entry rate... The broad range of indoor concentrations in the samples from Sweden, Canada and the United States is due primarily to differences in the rate of radon entry among dwellings. Thus, to understand the occurrence of high indoor concentrations, one must consider source materials and transport mechanisms by which radon enters houses.

As discussed in Chapter 5, sources of indoor radon include:

1. Soil. This is the main source of radon in most dwellings.
2. Building materials. These may be an important source in dwellings with low to moderate concentrations.
3. Domestic water. Water supplies with high levels of radon may be important in homes which utilize high quantities of water, particularly for bathing (showers).

Detailed studies of the factors which determine radon concentrations in homes have been carried out at the Lawrence Berkeley Laboratory (see e.g. Nero 1985a, 1986a). Wide variations were found in concentrations, due primarily to differences in the rate of entry of radon. Any correlations between radon concentrations and the ventilation rate were masked by the differences in the source strength.

Radon infiltration from the ground into a house is primarily due to pressure differences between the interior of the home and the soil. If the atmospheric pressure inside the home is lower than the pressure in the soil, flow into the house will be accelerated since air flows from high pressure to lower pressure. Pressure differences can arise due to wind action. Temperature differences between indoors and outdoors also affect the relative indoor and outdoor pressure. These temperature and pressure variations can produce a stack effect which sucks air in from the bottom of the structure where the interior pressure is lowered and out toward the top. The internal pressure in houses is usually less than the gas pressure in the soil. Internal pressure in houses, however, can be further lowered as the result of the operation of kitchen, bathroom and attic exhaust fans as well as by the use of fireplaces, furnaces and wood stoves, and clothes dryers. This further reduction in internal pressure will likely increase the entry of radon into the house.

B. Control techniques

1. Control of radon sources. Once the major source or sources of radon daughters has been established, consideration must be given to possible methods of control.

A study of a variety of radon control techniques especially directed towards reducing radon entry, was made for a group of houses in New York (Nitschke 1984). It was concluded that:

> ... houses with high radon levels cannot rely totally on ventilation... in very tight houses with low radon source strengths, whole house ventilation with heat recovery [utilizing air-to-air heat exchangers] may reduce radon to acceptable levels. Basement ventilation would be preferred to whole house ventilation if the radon source is in the basement (not in the water) and the basement is separated from other parts of the house. Sealing cracks and holes in basement walls and floors and ventilating the sub-slab can cause substantial reductions in indoor radon concentrations.

For homes with crawl spaces the most effective control technique involved ventilation of the crawl space.

The Pennsylvania Department of Environmental Resources (Department of Environmental Resources 1985) made a study of the routes by which radon enters houses and of mitigation measures. The study was for houses which were built with cinderblock basement walls. Not only is use of such walls a common construction practice, and therefore of specific interest, but it was concluded that the findings would be relevant to homes with other types of foundations whose walls "can also have cracks, penetrations, and other possible points of gas entry." The results of this study were used to establish an overall ranking of routes of entry of radon into basements.

For the cinderblock walls, which are usually more porous than ordinary concrete or stone, an important route of entry is through the wall itself, especially the bottom few feet. Other routes of entry, which are pertinent to many forms of basement construction, are exposed areas in the basement floor, such as drain sumps or unpaved areas, and cracks or other openings either in the floor or walls themselves and at joints. For an individual house, it was recommended that the basement first be inspected to identify likely sources and then that one proceed to the most cost-effective remedies. When these initial steps have been completed, their effectiveness should be checked by monitoring. If additional controls are indicated, it will usually be more complicated so that professional help should be utilized.

The Swedish Building Research Council funded experiments and evaluation of cost-effective radon control measures in existing buildings with high radon concentrations from infiltrating soil gas (Ericson 1984). It was found that a sub-floor pressure depression system, which keeps the air pressure in the ground lower than the air pressure in the basement, was an effective, low cost measure. This can best be accomplished by placing a pipe or series of pipes through the basement floor and then connecting these pipes to an exhaust fan discharging the contaminated air outdoors. The installation of a sub-floor depression system in 39 homes resulted in an 88% reduction in the average radon concentrations.

For another group of houses in this same study, the most significant source of radon infiltration from the soil was through a hole in the concrete surrounding the water pipes. In these homes, when the homes were sealed, the average radon concentration was reduced by 80%.

Radon emissions from building materials can be reduced via the application of surface coatings and barriers (Tartaglia 1984). The use of epoxy resin and polyester sheeting have been shown to be cost effective. The most effective method of controlling the problem of radon emissions from building materials is to avoid the use of products possessing a high radium content.

An Environmental Protection Agency report has recently summarized radon reduction methods (EPA 1986), listing the following techniques: natural and forced air ventilation, avoidance of house depressurization, sealing major sources and routes of entry, and ventilation of drain tiles, of hollow-block basement walls and of the soil underneath concrete slabs.

Additional natural ventilation can be provided via windows and vents. Outdoor air can also be introduced into the structure by fans or by using a forced air ventilation system with heat recovery. These systems reduce radon levels by the replacement and dilution of indoor air with outdoor air. House depressurization can be minimized by providing makeup air directly to appliances such as clothes dryers and heating systems which normally draw air from the house interior. The sealing of sources or of entry routes such as cracks in the structure has been successful in some cases, but the effectiveness of this approach has been described as being "extremely case specific." Exhaust ventillation systems for drain tiles, basement walls, or sub-slab soil have in some cases produced large reductions in radon concentrations, but the cost is relatively high. Further details of these approaches are presented in the EPA document (1986): *Radon Reduction Techniques for Detached Houses, Technical Guidance.*

2. Air cleaning techniques. If the above methods of control are inadequate or impractical and excessive levels of radon are detected inside the home, other methods of control utilizing passive or active radon reduction methods should be considered. A major passive removal process involves plateout or surface deposition (Jonassen 1984). In addition, radon daughters may be removed from the atmosphere by various active methods including filtration, electrostatic air cleaning and ion generation.

As discussed at greater length in Chapter 7, the radon daughters have a greater biological impact if they are inhaled in the form of unattached atoms, rather than as atoms attached to aerosol particles. The fraction of daughter atoms in the attached and unattached states is largely determined by the aerosol concentration in the atmosphere. Thus, cleaning measures which reduce the number of aerosol particles, such as those discussed below, may lead to a higher unattached fraction and may not be very effective in reducing the total radiation dose to the lung. More studies are necessary to determine their effectiveness.

Particulates can be removed from the environment by a filtration device made up of a suction source used to direct air through a mechanical filter (Sandia National Laboratories 1982). This system is designed to intercept and retain particles utilizing a filter matrix consisting of an agglomeration of fibers (fiber glass, cellulose, etc.) bonded together with a binder. Efficiency of collection depends in part on the pore size of the filter media. High efficiency particulate air (HEPA) filters are primarily used to remove small particles, below 0.3 micons diameter. If the atmosphere contains a significant number of large particles (greater than 10 microns in diameter) a coarse filter will usually precede the HEPA filter. While filtration can succeed in controlling radon daughters, a large portion of the daughters may remain in the unattached state so that there may be little reduction in the received dose.

Electrostatic air cleaners remove particulates from the atmosphere by establishing an electric field such that charged radon daughter products will move in the field and deposit on oppositely charged collection surfaces. The electric field is established by directing a high voltage (12KV, d-c) to a wire or plate electrode resulting in a charged corona. The corona releases positive ions which attach themselves to aerosols which in turn are deposited on oppositely charged collection electrodes. Electrostatic cleaners possess a relatively high efficiency for removal of aerosols in the respirable range. One drawback is that these devices liberate ozone, a toxic chemical.

Ion generators operate in a similar manner to the electostatic air cleaners, producing a charged corona. This corona emits ions which attach themselves to airborne particles. These charged particles are then attached to wall and ceiling surfaces. As with electrostatic air cleaners, ozone is also produced by ion generators.

C. Energy conservation and weatherizing

With the advent of energy conservation, tightening of older homes and construction of energy efficient homes has become common practice.

The design and construction of energy efficient homes varies widely. One type of energy efficient design, described by Walsh *et al.* (1984), was developed by the Arkansas Power and Light Company and is referred to as the "Arkansas Home." These homes are one-story structures with ventilated crawl spaces. They have R-38 insulation in the ceiling and R-19 insulation in the floors and walls. Double glazed windows occupy no more than 15% of the floor area. Insulated exterior doors are equipped with magnetic weatherstripping. The homes are constructed with extensive use of vapor barriers made of 6-mil polyethylene sheets, which cover almost all the surfaces through which air leakage might occur. Caulking is applied where electrical or plumbing pipes penetrate the structure. Although the radon problem was probably not a consideration in the design of these structures, some of the construction features will act to inhibit the entry of radon.

Sixty percent of the houses in the U.S. are more than 20 years old, while the replacement rate averages approximately 1.5% per year (Walsh 1984). It is obvious, then, that measures to make these older homes energy efficient will require a significant effort if energy conservation is to be successful. Weatherizing of the older structures will require weatherstripping and caulking of windows and doors as well as window and door frames. Any penetration of the outer building shell, such as by plumbing, must be sealed. Furthermore, kitchen and bathroom vents should be supplied with operable dampers. For complete energy conservation, the home should be adequately insulated. Insulation, however, does not appreciably tighten the home against air infiltration.

This movement toward energy conservation entails the lowering of building air infiltration rates with a concommitant reduction in ventilation, an important factor in the control of indoor pollution. We are, therefore, faced with a dilemma in that the tightening of homes will likely result in a increase in indoor pollution, including radon.

Theoretically, if the rate of radon entry is constant, then the indoor radon concentration should increase in direct proportion to the decrease in infiltration rate. Thus, if the number of air changes per hour is reduced from 1.0 to 0.5, the radon concentration should double, suggesting that energy conservation may well increase the indoor radon problem. In fact, this was an early concern (see, e.g., Budnitz 1979).

It appears, however, that in some instances the typical program of weatherstripping and caulking will reduce air infiltration rates by only a relatively small amount, averaging in the neighborhood of 10 to 20% (Nero 1986b).

On the other hand, a comparison of air changes per hour in 18 new energy efficient homes and 13 retrofit houses demonstrated that the average infiltration rate was slightly higher in the new homes (Walsh 1984). The average air changes per hour in new homes was 0.33 with a high of 0.92 and a low of 0.08, while the average in the retrofit homes was 0.30 with a range of 0.38 to 0.23. It would appear, therefore, that it is possible to tighten homes by retrofit so that the air infiltration rates approach those of newly constructed energy efficient homes.

In another study, the median average infiltration rate varied from 0.9 air changes per hour in old homes to 0.5 air changes per hour in newer homes. It was of interest to note that air infiltration rates monitored in a single home fluctuated with time from 0.1 to 1.1 air changes per hour (Perhac 1985).

A marked correlation between energy-efficient construction and radon concentration was reported by Fleischer and collaborators (1983), who found radon concentrations to be two to three times as high in a group of energy efficient homes as in a group of ordinary homes. Many of the energy efficient homes were solar homes with heat storage materials which release radon and could lead to high radon levels apart from any change in air exchange rates. On the other hand, the Bonneville Power Administration in a comparison of model conservation homes and other recently built homes found little significant difference in average radon concentrations (Reiland 1985), although here the model homes may have benefited from atypically tight basement construction which prevents radon entry.

In none of these studies were carefully matched houses used, identical in all respects other than ventilation conditions. In the absence of such a controlled comparison, or of controlled measurements of radon concentrations in a given group of houses before and after weatherization, it is not possible to give a reliable estimate of the effects of weatherization. Considerable further study will be required before the relationship between weatherization and radon concentration is well established.

D. Conclusions

For any given building, it is almost impossible to state precisely the routes of entry of radon into the structure. It is obvious, however, that the most critical source of radon entry into homes involves emissions from soil into building. Other radon sources such as building materials, water, and natural gas supplies, gain in importance when soil emissions are low. In the succeeding paragraphs, steps are indicated which can help minimize radon problems in existing and proposed buildings.

For existing homes, the first step is to determine the radon concentration and, if excessive, attempt to determine the sources. The structure should be examined for possible routes of entry. To reduce entry from the basement, one should seal all cracks and leaks, pour concrete on dirt floors and, if need be, install a ventilation system under the floor of the basement. If there is a crawl space, radon entry from it can be reduced by covering the ground with polyethylene sheeting or, as the most effective step, by ventilating the crawl space to the outdoors. If building materials are a major source, and their removal is not practical, alternatives are to seal them with a protective coating or apply a barrier.

For new buildings, the simplest step is to avoid building upon soil with a high radon emanation rate. In any event, radon levels will be minimized by utilizing a ventilated crawl space rather than a basement. In addition, it obviously helps to avoid the use of building materials with high radon emanation rates and the use of natural gas with unvented indoor appliances, if there is an unusually high radon concentration in the local gas supply. Furthermore, radon levels in the water supply should be checked. If high, controls, such as aeration, may be desirable.

During the next few years, a major nationwide effort would be desirable to better evaluate radon levels in various structures as well as to pinpoint more accurately sections of the country with high radon containing soils. The major questions confronting public health professionals concerns the human health effects attributable to the wide range of radon concentrations in the indoor environments, what standards should be adopted for "safe" radon exposures, and what are the most cost effective corrective measures that can reasonably be utilized for indoor environments exceeding this standard.

In implementing a program of remedial action, assuming Federal and State help, it would be well to concentrate on those homes which have particularly high radon levels. The cutoff level for defining "high levels" is, of course, somewhat arbitrary. A perspective on the priorities has been given in a Letter to the New York Times by leaders of the University of California group working in this area (Nero 1985b). In commenting on a contemplated EPA program which would require remedial action in about 5 million homes, the authors suggested that "a more probable program would be aimed at finding and fixing the million or so homes with levels more than five times the average." At present, we appear to be a considerable distance from undertaking either of these suggested programs.

As attempts proceed to set a radon standard, along with other indoor air standards, it should be kept in mind that there are likely to be other hazardous pollutants such as ozone, formaldehyde, and oxides of nitrogen, the combined effects of which might well be additive, or more important, synergistic. For example, electrostatic precipitators, utilized to clean indoor air, produce ozone. Borek *et al.* (1986) recently reported on studies of ozone exposure of hamster embryo cells and mouse fibroblast cells. Cell transformations were produced with ozone alone. Exposure of these cells to gamma radiation prior to ozone exposure produced increased transformations which were consistent with a synergistic reaction. The authors indicated that ozone acts alone as a carcinogen and synergistically with ionizing radiation as a cocarcinogen.

In some instances, the procedures necessary to control the risk of one hazard may not be compatible with reducing the radon risk. Individuals allergic to house dust would require an environment as free as is practical from the offending dust. Effective removal of the dust could increase the unattached fraction. All the most reason that investigations to determine the overall health hazards in homes be as comprehensive as possible in order to evaluate and, more importantly, to reduce the health risk of the total home environment where the average individual spends over 65% of each day and where the very young and the elderly spend more than 90% of each day.

Finally, despite all the work that has been accomplished to the present, it is well to keep in mind that strategies for controlling the indoor entry of radon are still in the research stage and some techniques indicated as useful in the past have since been shown to be ineffective.

References: Chapter 6

Borek, C., Zaider, M., Ong, A., Mason, H. and Witz, G., 1986, "Ozone acts alone and synergistically with ionizing radiation to induce in vitro neoplastic transformation," *Carcinogensis.* **7**, *1611.*

Budnitz R. J., Berk J. V., Hollowell C, D., Nazaroff W. W., Nero A. V. and Rosenfeld, A. H., 1979, "Human disease from radon exposures: the impact of energy conservation in residential buildings," *Energy and Buildings* **2**, 209.

Department of Environmental Resources, 1985, "General Remedial Action Details for Radon Gas Mitigation", Pennsylvania Department of Environmental Resources, Bureau of Radiation Protection.

Ericson S., Schmied H. E. and Clavensjo B., 1984, "Modified technology in new constructions, and cost effective remedial action in existing structures, to prevent infiltration of soil gas carrying radon," in: *Proceedings 3rd International Conference on Indoor Air Quality and Climate*, **5**, Aug. 20-24, 1984.

EPA (Environmental Protection Agency), 1986, *Radon Reduction Techniques for Detached Houses, Technical Guidance*, EPA/625/ 5-86/019 (June 1986).

Fleischer R. L., Mogro-Compero A. and Turner L. G., 1983, "Indoor radon levels in the northeastern U.S.: effects of energy-efficiency in homes," *Health Phys.* **45**, 407.

Jonassen N. and McLaughlin J.P., 1984, "Airborn radon daughters, behavior and removal," in: *Proceedings 3rd International Conference on Indoor Air Quality and Climate*, Vol. 2, Aug. 20-24, 1984.

Nazaroff W. W. and Nero A. V., Jr., 1984, "Transport of radon from soils into residences," in: *Proceedings 3rd International Conference on Indoor Air Quality and Climate*, Vol. 2, Aug. 20-24.

Nero A. V., Jr., 1985a, *Indoor Concentrations of Radon-222 and its Daughters: Sources, Range, and Environmental Influences*, Lawrence Berkeley Laboratory, Berkeley, CA 94720, LBL-19346.

______, 1986a, "The indoor radon story," *Technology Review*, January, p. 28.

Nero A. V., Jr. and Sextro R. G., 1985b, "A more practicable approach to reducing radon in houses," New York Times, Letter to Editor, November 8.

Nero A. V., Schwehr M. B., Nazaroff W. W. and Revzan K. L., 1986b, "Distribution of airborne radon-222 concentrations in U.S. homes," *Science*, **234**, 992.

Nitschke I. A., Wadach J. B., Clarke W. A., Traynor G. W., Adams G. P. and Rizzuto J. E., 1984, "A detailed study of inexpensive radon control techniques in New York State houses," in: *Proceedings 3rd International Conference on Indoor Air Quality and Climate*, Vol. 5, Aug. 20-24.

Perhac R. M., 1985, "Indoor Air Quality, Electric Utility Concerns," in *Indoor Air and Health* (Lewis Publishing Company).

Reiland P., McKinstry M. and Thor P., 1985, "Preliminary radon testing results for the residential standards demonstration program", Bonneville Power Administration, Department of Energy (August).

SNL (Sandia National Laboratories, 1982, *Indoor Air Quality Handbook*.

Tartaglia M., Di Nardi S. R. and Ludwig J., 1984, "Radon and its progeny in the indoor environment," *Journal of Environmental Health*, **47**, No. 2, Sept.- Oct.

Walsh P. J., Dudney C. S. and Copenhaver E. D., 1984, *Indoor Air Quality* (Cleveland: Chemical Rubber Co. Press).

Chapter 7

DOSIMETRY MODELS

Maurice A. Robkin

A. Geometry of the Respiratory System

1. Configuration of the airways. In this Chapter, the radiation dose delivered to the lung by inhaled radon daughters will be considered. Only the effects of the short-lived daughters are included (i.e., lead-210, with a 22-year half-life will be ignored). For this discussion, it is necessary to consider the geometry of the respiratory system. A common description, which will be used in this discussion, divides the respiratory system into three parts. The uppermost part consisting of the nose, mouth, throat, pharynx and larynx is denoted as the naso-pharyngeal (N-P) region. In this region inspired and expired air are moving at their greatest speed and air flow is most likely to be turbulent. The particles carried in the inhaled air have different depositional behavior in the different parts of the lung depending on whether the air flow is turbulent or laminar (i.e., smooth). Laminar flow is typical of low airspeeds.

The next part of the respiratory tract is composed of the larger airways. This part begins at the trachea which divides into two large airways. Each airway further divides, and these divisions continue until their diameters and lengths become quite small. Each division can be numbered starting at the division of the trachea. The section of airway between two successive divisions is called a "generation". Thus, if we refer to the trachea as generation number 0, there are two airways in generation number 1.

Each airway does not necessarily give rise to only two smaller airways. In the right lung, the first generation airway gives rise to three second generation airways while, in the left lung, the first generation airway gives rise to two second generation airways. However, for simplicity, one of the lung models discussed below ignores this complication and assumes a regular symmetric splitting at every branching.

Small groups of generations are given descriptive names. The two airways of generation 1 are called the "main-stem" or "primary" bronchi, generations 2 and 3 are called the "lobar bronchi" or "secondary bronchioles", generations 4 to 6 are called the "segmental bronchi" or "tertiary bronchioles" and so forth until the airways of generations 10 to 16 which are called the "terminal bronchioles". The part of the respiratory tree from generation 0 to generation 16 is referred to as the tracheo-bronchial (T-B) region. By the time the airflow has been divided several times, the flow speed has become quite slow and the flow is basically laminar in the remainder of the lung.

Beyond generation 16, the airway continues to divide until finally the smallest parts of the lung are reached. These are the alveoli where gas exchange with the blood is carried out. The part of the lung beyond generation 16 is called the pulmonary (P) region. The respiratory system is illustrated in Figure 7-1.

2. Description of the surface wall of the airways. The surface wall of the airways in the T-B region is made up of three parts (see Figure 7-2). The innermost layer furthest from the lumen of the airway is the epithelium. The epithelium is composed of an inner layer containing basal cells that divide to produce the cells of the outer layer. The outer layer of the epithelium is mostly made up of ciliated and goblet cells. The goblet cells are also thought to be capable of cell division (McDowell 1978a). The middle part of the airway wall is a layer of cilia. The cilia are in constant motion and serve to sweep foreign material trapped in the mucus. The mucus makes up the outermost double layer of the wall adjacent to the airway. The inner part of the mucus and the cilia occupy the same layer. The inner part of the mucus is a serous (watery) layer in which the cilia beat. The outer part is a viscous layer. This outermost layer of the mucus is propelled up the respiratory tree by the tips of the beating cilia and the trapped foreign particles are carried to the throat where they are swallowed and eliminated via the gut.

The airway surface of the pulmonary region is unciliated and has no mucus sheath. Special cells, called phagocytes, are mobilized to the surface of the P region where they engulf small foreign particles and transport them up the respiratory tree to the mucus transfer system of the T-B region. In either region, radioactivity may also be cleared by desorption or dissolving of the particles and the transfer of the activity to the blood across the blood-tissue interface or by transfer to the lymphatic system (TGLD 1966).

The target cells usually considered to be at risk for alpha-particle induced lung cancer are the dividing basal cells of the epithelium of the walls of the lung airways (see Figure 7-2). Other cells that occur in the epithelium, some of which can divide, are also implicated but to a lesser extent (McDowell 1978b). The most common location for the appearance of lung cancer seems to be in the epithelium of the segmental bronchi about at generation 4, where the calculated dose is greatest (Altshuler 1964; Harley 1972, 1981, 1982; BPA 1984). After cell division, some of these cells become the highly specialized ciliated cells which do not divide. These non-dividing cells age and are lost from the surface layer and are not at risk for becoming cancerous. Normally, the rate of cell division of the progenitor cells is under strict regulation so that the number of new non-dividing cells formed is just balanced by the number of aged cells lost. Irradiation of dividing cells can cause them to lose regulatory control so that they grow in an uncontrolled manner forming a tumor, i.e., lung cancer.

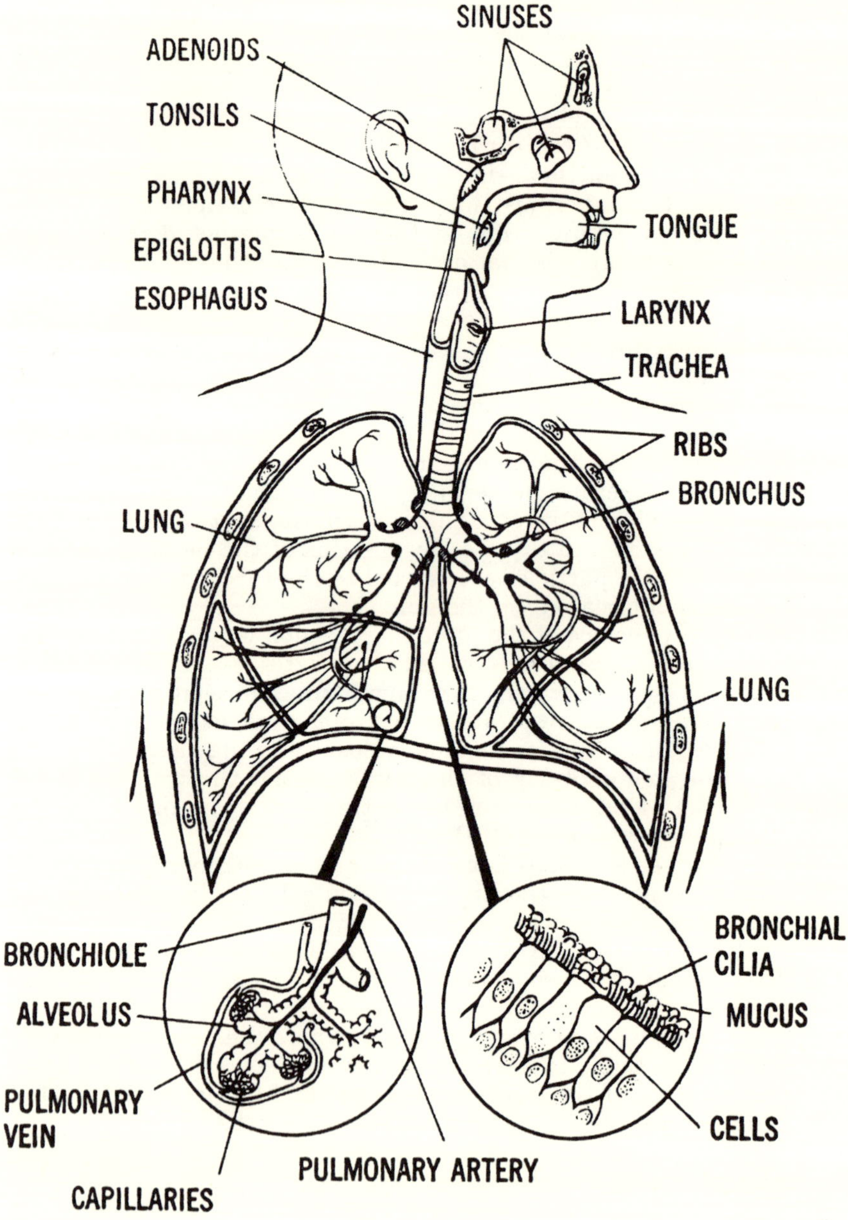

FIG. 7-1. Diagram of the respiratory system. The inserts show enlarged pictures of an alveolus (gas exchange area) and of a section of bronchial tube wall. [Reprinted, with permission, from *Occupational Lung Diseases, An Introduction*, American Lung Association, 1979, p. 10.]

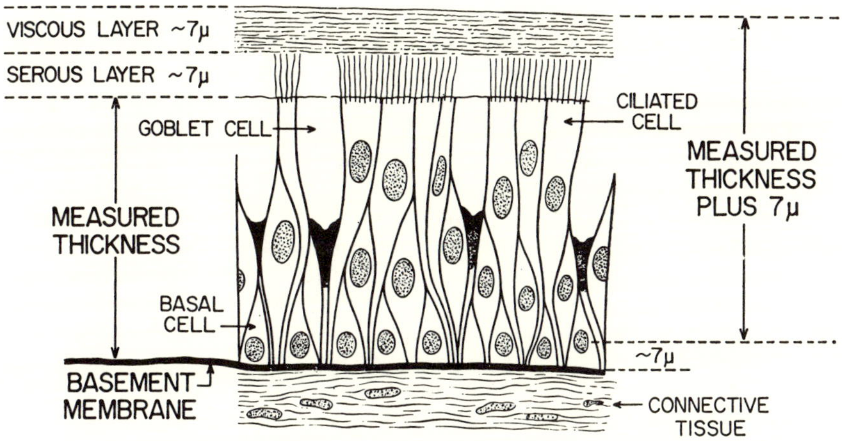

FIG. 7-2. Schematic illustration of the bronchial epithelium and double-layered mucus, with estimates of relevant dimensions. [Reprinted, with permission, from Altshuler *et al. Health Physics*, Pergamon Press (Altshuler 1964, Figure 3).]

To define the irradiation geometry, the thicknesses of the different layers must be specified. These thicknesses vary both from one lung generation to another and within a generation. There are differences among men, women and children, and variability from person to person. As a result, it is not possible to define an unique geometry, and difficult to establish an average geometry. Even averaging the dosimetry over the entire lung of a "standard adult male" requires an extremely detailed and complex computation.

Gastineau *et al.* (1972) have made measurements of epithelial dimensions in various generations of the lung using surgical specimens. Their results are reported without specification of age or sex of the individuals providing the specimens. If it is assumed that the length of the cilia defines the thickness of the serous layer of the mucus and that the serous and the viscous mucus layers have the same thickness, then in the segmental bronchi their measurements give the following ranges of thicknesses: 2.5 to 10 microns each for the cilia and viscous mucus and 10 to 90 microns for the epithelium, exclusive of the basal cell layer which ranges from 5 to 10 microns.

B. General calculational approach

In order to calculate the dose delivered to the basal cells from alpha particles produced by the short-lived radon daughters which are trapped in the mucus, a sequence of computations is required. These include the determination of the concentration of short lived daughters of radon in the air being breathed, the determination of the amount of activity on the airway surfaces of the lung due to deposition and clearance, and the calculation of the actual energy delivered to the sensitive tissues. For this calculation, it is necessary to specify the deposition and removal rates of the daughters. The details of these are discussed in Chapter 7, Section D. Given these rates, and the rates of radioactive decay for each daughter, the rate of production of alpha-particles can be computed.

The alpha-particles are distributed in the mucous layer where they form an irradiation source. In some models, the source is assumed to be confined to the surface of the mucus. In others, the source is assumed to be distributed throughout the mucus. In some models, some of the radioactive radon daughters are cleared by being dissolved from their carrier particles and transferred through the epithelium to the blood. Thus, the source is allowed to move from the mucus to the epithelium. The importance of this to the dosimetry depends on how long it takes for the activity to pass through the epithelium and the amount of activity which is cleared by this route.

The atmospheric aerosol typically ranges in size from several hundredths of a micron to several microns. As discussed in Chapter 2, the radioactive daughters of radon, when they are formed in the air, are highly ionized and very reactive. They can react chemically with components of the air or form small ionized agglomerates with water vapor or both. These ions have a size range of the order of one-thousandth the size of the atmospheric aerosol. Many of these attach to the atmospheric aerosol. Depending on the size distribution of the atmospheric aerosol, on the fraction of the activity which is attached to the aerosol and the fraction which occurs as unattached small ions, the distribution of deposition of the radon daughters in the various regions of the lung is established.

C. Models of the lung

The ramifications of the respiratory tree are usually described in terms of the number of splits of the airway, or bifurcations, which have occurred between the trachea and the particular location. There are two models of the respiratory tree commonly used for lung dosimetry. These are the Weibel "A" model (1963) which treats the lung as a single unit with regular symmetric bifurcations and the Yeh-Schum model in which the lung is either divided into five lobes, each of which has its own subdivision pattern, or is averaged into a representative whole lung (Yeh 1980).

With the definition of "generation" given in Section A as the region between two successive bifurcations, each generation is given an index according to the number of bifurcations between it and the trachea. The trachea is denoted as generation 0 in the Weibel-A model or generation 1 in the Yeh-Schum model.

The airway diameters of a given generation are slightly larger in the Yeh-Schum model which results in the deposition probabilities based on this model being slightly different from those based on the Weibel-A model. These differences are the result of measurement differences obtained by Yeh and Schum (1983) from casts of the human lung relative to the measurements obtained by Weibel (1963) from casts and histologic preparations.

These two lung models form the basis of two of the most commonly used lung dosimetry models, the Jacobi-Eisfeld (J-E) model (1980) which uses the Weibel-A lung model and the James-Birchall (J-B) model (1980) which considers both whole-lung models. An excellent description of these models and the dosimetry derived from them is provided in an extensive report of an expert group published by the Nuclear Energy Agency of the Organization for Economic Cooperation and Development (NEA 1983).

For the purposes of illustrating some of these concepts, a simple dosimetric model based on treating the T-B region as a single compartment with average parameters is discussed in Appendix A.

D. Deposition of radon daughters in the lung

1. Deposition of the unattached fraction. The atoms or ions which comprise the unattached fraction of the radon daughters deposit by simple diffusion to the walls of the respiratory tract. At a breathing rate of 750 liters per hour, a value typical for adults engaged in no more than light activity, about 65% of the inhaled unattached fraction is removed in the nose (George 1969). Of the remainder, the deposition probability per unit area on the inner surfaces of the respiratory tree is much greater in the larger upper airways.

Deposition of unattached atoms occurs higher in the respiratory tract in the Jacobi-Eisfeld model than in the James-Birchall model. This is due to the assumption in the J-E model that deposition in the upper airways is enhanced over that for laminar flow due to the turbulence arising from the disturbing effect of the bifurcations on the flow pattern. In the J-E model, deposition falls off rapidly in generations deeper than generation 5. In the J-B model, the rapid fall off occurs after generation 10. Electric forces due to a charge on the free ions may play some role in the rapid deposition of the unattached fraction in the upper parts of the respiratory tract. In both models, there is essentially no deposition of the unattached fraction in the pulmonary region.

Because the two lung models use different dimensions for each generation of airway in the lung, the deposition probabilities obtained for the various deposition mechanisms are different, which leads to differences in the lung dosimetry for inhaled radon daughters. In addition, the breathing rate and dimensions of the airways of the lung are a function of age, so that there is also a dependence of deposition on the age of the individual which takes substantially different form in the two models. For example, in the J-E model for one-year-olds, there is no deposition of the unattached fraction beyond generation 0 while in the J-B model, the deposition probability per unit area for one-year-olds is rather broadly spread from generation 0 to about generation 8, after which it rapidly falls off (NEA 1983).

2. Deposition of the attached fraction. Deposition in the lung of the fraction of the radon daughters attached to the aerosol is mainly the result of three processes. These are diffusion, impaction, and sedimentation. Diffusive deposition is the result of the random Brownian motion of small particles bringing them into contact with the surface of the lung. Impactive deposition is the result of inertia which prevents a particle from negotiating an abrupt change in the direction of an air streamline following a bifurcation in an airway so that it strikes the airway surface. Sedimentative deposition is the result of gravity acting on a particle causing it to settle out onto the airway surface.

The probability of the deposition of attached atoms per unit area of respiratory tract surface is a strong function of the size distribution of the atmospheric aerosol to which they are attached. In the T-B region, both the J-E model and the J-B model predict a maximum in the deposition probability per unit area in generation 3 (NEA 1983, Weibel notation) for the aerosol size distributions characteristic of indoor air, namely those with mean diameters in the neighborhood of 0.1 microns to 0.25 microns. For atmospheric aerosol sized distributions characteristic of indoor aerosols, both the J-E and the J-B models give quite similar results for deposition in generations 5 through 15. In both models, the deposition probabilities increase with increasing breathing rate in every lung generation.

3. Removal mechanisms. The deposition represents a source term for the activity balance in the lung. The loss terms include physical decay and biological removal. The J-E model and the J-B model use different retention times based on different explicit treatment of similar removal mechanisms. These mechanisms include transfer of particles upwards in the moving mucous layer from one generation to the next and desorption from or dissolving of the carrier aerosol, followed by diffusion of the released radionuclides into the epithelium where they can either remain or transfer into the bloodstream and be carried away.

The muco-ciliary escalator which carries deposited particles up the respiratory tree to the throat causes each generation of airway to be exposed to particles originally deposited more deeply in the lung. The time it takes to transfer particles from one generation to the next by this mechanism is, in the upper generations, relatively short compared to the rate of radioactive decay. In the lower generations, deeper than generation 6, different models have very different transit times ranging from a few minutes to no transfer at all (NCRP 1984).

The importance of these variations depends on the fraction of the activity cleared by this route. In the J-E and J-B compartment transfer models, every generation is represented by a compartment and transfer between lung generation compartments is accounted for explicitly (NEA 1983).

In the NEA description of the lung (1983), the overall bronchial clearance time for muco-ciliary clearance is much longer than the radioactive half-lives of the short-lived radon daughters. In the ICRP lung model (1972), 95% of the deposited activity is cleared to the blood with a half-life of about 14 minutes while 5% of the activity is cleared via the muco-ciliary escalator with a half-life of 288 minutes. However in the models used both by the NEA experts group (1983) and the National Council on Radiation Protection and Measurements Task Group (1984), it appears that for the most part clearance to the blood takes a long time relative to the radioactive decay times.

For a first approximation, clearance mechanisms may be neglected and only radioactive decay used as a loss term. For accurate estimations of dosimetry, however, the biological clearance mechanisms should be taken into account. The J-E and J-B lung dosimetry models, discussed below, account for biological clearance.

E. Calculation of the radiation dose

The daughter activity in the lung reaches a balance such that the rate of loss equals the rate of deposition. Once the balance amount of daughter activity in the lung is established, the amount of lung tissue irradiated is determined. The basal cells of the T-B region lie beneath a layer of mucus, cilia and epithelium, at a typical depth of the order of 50 microns (see Chapter 7, Section D-2). The approaches to the location of the sensitive basal cells taken by Jacobi and Eisfeld and by James and Birchall are somewhat different (NEA 1983). In the J-E model, the depth of the basal cell layer is assumed to continually decrease with increasing generation index. The dose to the basal cells is computed on the basis of the mean depth in each generation. In the J-B model, the basal cells are taken to lie at the distribution of depths given by Gastineau *et al.* (1972). The shielding effect of the mucous layer and the serous layer containing the cilia are combined with the distributions of the deposited radon daughters and the depths of the target basal stem cells to establish the irradiation geometry for the dose calculation due to irradiation by alpha particles. The 6.00-MeV alpha particle of polonium-218 has a range in tissue of about 47 microns. The 7.69-MeV alpha particle of polonium-214 has a range in tissue of about 71 microns (Lea 1947). In both the J-E and the J-B models, some alpha particles of both energies can reach some target cells in every lung generation, but the amount of shielding provided by the overlying cells and mucus differs. The dose per alpha particle is larger in the J-B model than in the J-E model (NEA 1983).

Recognizing the enhanced contribution to the dose per working level of the unattached fraction, the resulting dose per unit exposure can be expressed in the form:

$$\text{tracheobronchial region: } D_{T-B} = f_p \times D^u_{T-B} + (1 - f_p) \times D^a_{T-B}$$
$$\text{pulmonary region: } D_P = (1 - f_p) \times D^a_P$$

where D denotes the dose (in rads per WLM or grays per joule) and f_p is the fraction of the total potential alpha energy that is unattached. The unattached fraction is usually taken to be made up entirely of RaA. D^u_{T-B} is the mean regional dose for unattached daughters in the T-B region, D^a_{T-B} is the mean regional dose in the T-B region for the attached fraction, and D^a_P is the mean regional dose in the P region for the attached fraction. The equation for D_P reflects the fact that all of the unattached activity is removed in the N-P and T-B regions.

Tables 7-1 gives values of the parameters for the above equations in the J-E and J-B models for adult members of the public breathing at a rate of 750 liters of air per hour. These results are presented in terms of the regional lung dose in rads/WLM as a function of the unattached fraction, f_p, for the cases of aerosols with activity median diameters (AMD) of 0.1 and 0.2 microns. As can be seen from Table 7-1, the dose predicted by the J-B model increases much more rapidly with the unattached fraction than does the dose based on the J-E model.

TABLE 7-1. *Conversion factors for calculating the regional lung dose (rad per WLM) for indoor radon exposures, in models presented in Nuclear Energy Agency study. (Results are given for the tracheo-bronchial (T-B) and pulmonary (P) regions for two values of the activity median diameter (AMD).)*

Dosimetric Model	AMD=0.1 micron T-B Region	AMD=0.1 micron P Region	AMD=0.2 micron T-B Region	AMD=0.2 micron P Region
J-E	$0.53 + 1.5f_p$	$0.13(1-f_p)$	$0.29 + 1.7f_p$	$0.09(1-f_p)$
J-B	$0.50 + 6.2f_p$	$0.05(1-f_p)$	$0.28 + 6.4f_p$	$0.03(1-f_p)$

Reference: NEA 1983, Table 2.9.
Assumptions: Nose breathing; 750 liter per hour breathing rate.

There will be a wide spread in values of the median aerosol size and the magnitude of the unattached fraction from house to house, depending in part upon ventilation conditions. The NEA experts group assumes that typical indoor environments are best characterized by "low" to "moderate" ventilation rates (in the neighborhood of 0.5

air changes per hour), corresponding to parameter values of about: AMD = 0.17 microns and f_P = 0.02. This leads to dose conversion factors, averaged between the J-E and J-B models, of approximately (NEA 1983, Table 2.11):

$$D_{T-B} = 400 \text{ mrad/WLM}$$
$$D_P = 50 \text{ mrad/WLM}$$

Somewhat different parameters are assumed for the dosimetry model of NCRP Report No. 78 (1984). The parameters adopted in this Report are: AMD = 0.125 microns, Rn/RaA/RaB/RaC = 1.0/0.9/0.7/0.7 and the ratio of the unattached RaA activity to the radon activity, (f_A), is 0.07. Other unattached activity is neglected in the NCRP model. For these parameter values, the unattached fraction of the potential alpha energy, f_P, is 0.01. In addition, an average breathing rate of 930 liters per hour is assumed, rather than the 750 liters per hour assumed in NEA 1983. Expressed for the adult male (intermediate between children and adult females), these assumptions lead in the NCRP calculation to the tracheo-bronchial dose conversion factor (NCRP 1984):

$$D_{T-B} = 710 \text{ mrad/WLM}$$

Table 7-2 evaluates the weighted whole body dose equivalent for the NEA models assuming a quality factor of 20 for alpha particles and a weighting factor of 0.06 for each lung region, based on sharing the 0.12 lung weighting factor for the whole lung (ICRP 1979) between the T-B and P regions (NEA 1983). A more extensive discussion of the conversion of the regional physical doses (rads) to the whole body equivalent dose commitment (rems) is given in Chapter 10.

TABLE 7-2. *Conversion factors for calculating the whole-body effective dose equivalent (rem per WLM) for indoor radon daughter exposures, in models presented in NEA 1983. (Results are given for two values of the activity median diameter (AMD)).*

Dosimetric Model	AMD=0.1 micron	AMD=0.2 micron
J-E	$0.79 + 1.6f_p$	$0.45 + 1.9f_p$
J-B	$0.66 + 7.4f_p$	$0.37 + 7.6f_p$

Reference: NEA 1983, Table 2.6.
Assumptions: Quality Factor of 20 for alpha-rays; whole body equivalent weighting factors of 0.06 for T-B and P lung regions.

APPENDIX A

In this Appendix, a simple model for lung dosimetry will be developed and a sample calculation done. The model to be evaluated is as follows. The doses will be calculated for one working level month (WLM) based on breathing an atmosphere containing one working level (WL) of short lived radon daughters for one working month of 170 hours. The atmosphere contains RaA, RaB, and RaC, with activity measured in pCi/ℓ. The activities of these isotopes are in the ratio, a:b:c, where a, b, and c represent the fraction of the equilibrium activities which are present. The notation *a, b,* and *c* is equivalent to the notation $c_A/c_{Rn}, c_B/c_{Rn}$, and c_c/c_{Rn} of Chapter 2.

The half-lives of RaA, RaB, and RaC, in minutes, are 3.11, 26.8, and 19.8 respectively. With these values, one pCi of each of these isotopes contains 9.96, 85.8 and 63.4 atoms respectively. As discussed in Chapter 2, every atom of RaA, as it and its daughters decay, will eventually generate 13.69 MeV of alpha energy and every atom of RaB or RaC will generate 7.69 MeV of alpha energy. Recall that the short lived daughters in equilibrium with 100 pCi/ℓ of radon will generate 1 WL of alpha-particle energy. If the radon has a specific activity of 100 pCi/ℓ and the daughters have specific activities of 100a, 100b, and 100c pCi/ℓ, then they will generate F working levels where (see Table 2-3).

$$F = 0.106a + 0.514b + 0.380c$$

Conversely, if the short-lived daughters are generating 1 WL, then the radon concentration must be 100/F pCi/ℓ and the daughter activities are 100a/F, 100b/F, and 100c/F, respectively.

The RaA activity is usually divided into two parts. One part is made up of small ions which are unattached to the atmospheric aerosol, (f_A) The other part is RaA activity which is attached to the particles of the atmospheric aerosol, (a-f_A). Although both RaB and RaC may have unattached activity, the amounts are usually small and they will be ignored and all of their activity considered to be attached. Thus, the attached activities of RaA, RaB, and RaC have specific activities of 100 (a-f_A)/F, 100b/F and 100c/F, respectively.

The airways of the tracheo-bronchial region are made up of a set of connected tubes. In the Weibel-A model of the lung (1963), each generation of airways is formed by a symmetric splitting of the airway of the previous generation. For example, generations 4, 5 and 6, which are denoted as the segmental bronchi, contain 16, 32 and 64 airways respectively.

Activity which is breathed deposits on the inner surface of the airways. Deposition probabilities for the unattached and attached activities in each generation for the Weibel-A lung model are given in graphical form in the report of the NEA experts panel (1983). They consider that the tracheo-bronchial region is composed of the trachea plus the airways resulting from 15 splittings (NEA 1983).

Assuming a mean aerodynamic diameter of 0.1 microns for the

atmospheric aerosol, using the NEA values for the breathing rate appropriate for the population at large (0.75 cubic meters per minute) and the dimensions of the Weibel-A lung model, and summing the depositions from generation 0 through generation 15 over their total surface areas, the fraction, U, of the unattached activity and the fraction, D, of the attached activity which deposit in the tracheo-bronchial region are approximately $U = 0.5$ and $D = 0.1$.

RaA decays into RaB which decays into RaC and all of these activities may be removed from the lung by biological clearance processes. There are various estimates given for biological clearance of the activity of the radon daughters deposited on the surface of the airways (NEA 1983, NCRP 1984). The clearance times range in value from several minutes to several hours for clearance to the blood and over similar values for clearance via the muco-ciliary escalator of the T-B region. Neglecting the biological clearance gives a conservative (higher dose) estimate of the dosimetry. Assuming that the specific activities of the radon daughters in the atmosphere and the breathing rates remain constant for long periods of time, the surface activity in the lung will build up to a value given by the balance between losses and deposition.

One picocurie of activity corresponds to 2.22 disintegrations per minute. One working level of short-lived radon daughters at an equilibrium level, F, is made up of 222x/F disintegrations per minute per litre for each daughter, where "x" takes on the value of f_A, a-f_A, b, and c for RaA*, RaA, RaB, and RaC respectively. The production rate of 6.00-MeV alpha particles in the T-B region is equal to the decay rate of RaA there. The production rate of 7.69-MeV alpha particles per unit area of airway surface is equal to the decay rate of RaC.

The equilibrium activities, in alpha particles per minute per WL of daughter activity in the air, can be expressed as:

$$S_a = (222/\ln 2) \times (V/F) \times [(U - D) \times f_A + D \times a] \times T_a$$

$$S_c = (222/\ln 2) \times (V/F) \times \{[(U - D) f_A T_A + (aT_A + bT_B + cT_c) D\}$$

where T_A, T_B and T_c are the half-lives of RaA, RaB, and RaC in minutes.

Various estimates have been given for the breathing rate appropriate for the population at large. For the NEA experts group's value of 0.75 cubic meters per hour, $V = 12.5$ liters per minute. A representative distribution of daughter activities for indoor environments a:b:c: = 0.85:0.55:0.35 with $f_A = 0.07$. For these values, the equilibrium activities for the T-B region are:

$$S_a = 9,800 f_A + 2,100 \text{ alphas per minute}$$

$$S_c = 9,800 f_A + 19,000 \text{ alphas per minute}$$

The estimated tissue weight of the tracheo-bronchial region (generation 0-15) is about 46 grams for the adult (NCRP 1975). Assuming that the alpha particles generated by the radon daughters deposited in the T-B region deposit all of their energy there, and noting that there are 1.6×10^{-8} gram-rad per MeV and that one working month of 170 hours contains 10,200 minutes, the doses generated by the decay of RaA and RaC are:

$$D_a = 1.6 \times 10^{-8}\,\text{gm-rad/MeV} \times 10,200\,\text{min} \times (6.00\,\text{MeV}/46\,\text{g}) \times S_a$$

$$D_c = 1.6 \times 10^{-8}\,\text{gm-rad/MeV} \times 10,200\,\text{min} \times (7.69\,\text{MeV}/46\,\text{g}) \times S_c$$

On the basis of the above model, the regional dose per working-level-month in the tracheo-bronchial region is

$$D = 0.48\text{f}_\text{A} + 0.56\ \text{rads/WLM}$$

The dose per WLM given by the NEA experts group for the T-B region for adult members of the public is, for the Weibel-A model as used by Jacobi and Eisfeld (Table 7-2),

$$D = 1.5\text{f}_\text{p} + 0.53\ \text{rads/WLM}$$

where f_A, f_B, f_C and f_p are related by

$$\text{f}_\text{p} = \frac{0.106\text{f}_\text{A} + 0.514\text{f}_\text{B} + 0.380\text{f}_\text{C}}{\text{F}}$$

For the assumed daughter activity distribution, $\text{f}_\text{A} = 0.07$, $\text{f}_\text{B} = \text{f}_\text{C} = 0$ and $F = 0.51$. Thus, $\text{f}_\text{p} = 0.015$. For these values, the T-B dose per WLM is 590 mrad for this simple model and 550 mrad for the NEA model.

In many environments, the unattached fraction of RaB is about 10% of that of RaA. For this case, and with the above values, $\text{f}_\text{p} = 0.022$. This large change has a small effect on the result of the NEA model and practically no effect on the result of the simple model. The NEA result becomes 560 mrad per WLM.

Considering the simplicity of the model in this Appendix, and noting that biological clearance was neglected which results in an overestimate of the doses, the agreement is remarkably good.

It may be noted that the deposition rate per unit mass of epithelium irradiated is a maximum in generations 3 and 4, where it is substantially above the average for the T-B region. This results in a much higher dose in these generations than in the other generations of the T-B region, which is suggestive of why lung cancers more frequently develop there.

References: Chapter 7

Altshuler B., Nelson N. and Kuschner M., 1964, "Estimation of lung tissue doses from the inhalation of radon and daughters," *Health Phys.* **10**, 1137.

BPA (Bonneville Power Administration), 1984, *The Expanded Residential Weatherization Program,* Vol. 1, App. F, F1.

Gastineau R. M., Walsh P. J. and Underwood N., 1972, "Thickness of bronchial epithelium with relation to exposure to radon," *Health Phys.* **23**, 860.

George A. and Breslin A. J., 1969, "Deposition of radon daughters in humans exposed to uranium mine atmospheres," *Health Phys.* **17** 115.

Harley N. H. and Pasternack B. S., 1972, "Alpha absorption measurements applied to lung dose from radon daughters," *Health Phys.* **23**, 771.

_____, 1981, "A model for predicting lung cancer risks induced by environmental levels of radon daughters," *Health Phys.* **40**, 307.

_____, 1982, "Environmental radon daughter alpha dose factors in a five-lobed human lung," *Health Phys.* **42**, 789.

ICRP (International Commission on Radiological Protection), 1972, "The Metabolism of Compounds of Plutonium and other Actinides," *ICRP Publication 19* (Oxford: Pergamon Press).

_____, 1979, " Limits for Intakes of Radionuclides by Workers," ICRP Publication 30, *Annals of ICRP* **2**, No. 3/4.

Jacobi W. and Eisfeld K., 1980, "Dose to tissues and effective dose equivalent by inhalation of radon-222, radon-220 and their short-lived daughters," GSF Report S-626 as quoted in NEA 1983.

James A. C, Greenhalgh J. R. and Birchall A., 1980, "A Dosimetric Model for Tissues of the Human Respiratory Tract at Risk from Inhaled Radon and Thoron Daughters," in: *Radiation Protection: A Systematic Approach to Safety*, Vol. 2, Proceedings of the 5th Congress of the International Radiation Protection Society, Jerusalem, March (Oxford: Pergamon), pp. 1045-1048.

Lea, D. E., 1947, *Actions of Radiations on Living Cells* (Cambridge: Cambridge University Press).

McDowell E. M., Barrett L. A., Glavin F., Harris C. C. and Trump B. F., 1978a, "The respiratory epithelium. I. Human bronchus," *J. Natl. Cancer Inst.* **61**, 539.

McDowell E. M., McLaughlin J. S., Merenyi D. K., Kiefler R. F., Harris C. C. and Trump B. F., 1978b, "The respiratory epithelium. V. Histogenesis of lung carcinoma in the human," *J. Natl. Cancer Inst.* **61**, 587.

NCRP (National Council on Radiation Protection and Measurements), 1984, "Evaluation of Occupational and Environmental Exposures to Radon and Radon Daughters in the United States," *NCRP Report No. 78* (Bethesda, MD: NCRP).

NEA (Nuclear Energy Agency, OECD), 1983, "Dosimetry Aspects of Exposure to Radon and Thoron Daughter Products," *Report by*

a Group of Experts (Paris: OECD).
TGLD (Task Group on Lung Dynamics), 1966, "Deposition and retention models for internal dosimetry of the human respiratory tract," *Health Phys.* **12**, 173.
Weibel E. R., 1963, *Morphometry of the Human Lung* (Berlin: Springer-Verlag).
Yeh H-C. and Schum G. M., 1980, "Models of human lung airways and their application to inhaled particle deposition," *Bull. Math. Biol.* **42**, 461.

Chapter 8

OBSERVATIONS OF LUNG CANCER: EVIDENCE RELATING LUNG CANCER TO RADON EXPOSURE

Kenneth L. Jackson, Joseph P. Geraci, and David Bodansky

A. Incidence of lung and other cancers in the general population

Lung cancer develops slowly over a period of years and unfortunately often shows no symptoms until late in its course. By the time symptoms cause an individual to seek medical care, lung cancer has frequently spread so that only about 25% of those with the disease are candidates for surgery. Although some lengthening of useful survival of individuals not successfully treated by surgery has been obtained with radiotherapy and chemotherapy, median survival time remains low, less than one year. Prevention is, therefore very important in the control of lung cancer.

Estimates of cancer incidence for various types of cancer have been tabulated by the American Cancer Society (ACS 1987). Of the 965,000 new cancer cases estimated in the United States for 1987, 150,000 are lung cancer. Only 13% of these lung cancer patients will survive five years or more. This low survival rate is primarily due to spread of the tumor (i.e., invasion of adjacent tissues and/or spread to other parts of the body). Of 483,000 estimated cancer deaths in 1987, 136,000 are from lung cancer. As is the case with most other cancers, lung cancer death rates increase with age.

The incidence of lung cancer varies by country, being intermediate for the United States. For example, the lung cancer rate per 100,000 males in 1980-81 was 72 in the United States, while in Japan it was only 34 and in Scotland was as high as 110 (ACS 1986). Lung cancer incidence in the United States has been increasing rapidly since 1930, particularly in women who smoke, whereas cancers of the liver, stomach, and uterus have been decreasing. Breast, colon and rectum, prostate, and pancreatic cancers, and leukemia have more or less stabilized (see Figure 8-1). Lung cancer has been the largest single source of cancer deaths for a number of years and now accounts for over one-quarter of the total cancer mortality. It is estimated that it has now become the major single killer even for women, surpassing breast cancer. The rise is attributed to cigarette smoking. The American Cancer Society estimates that smoking is responsible for 83% of all lung cancers (ACS 1987).

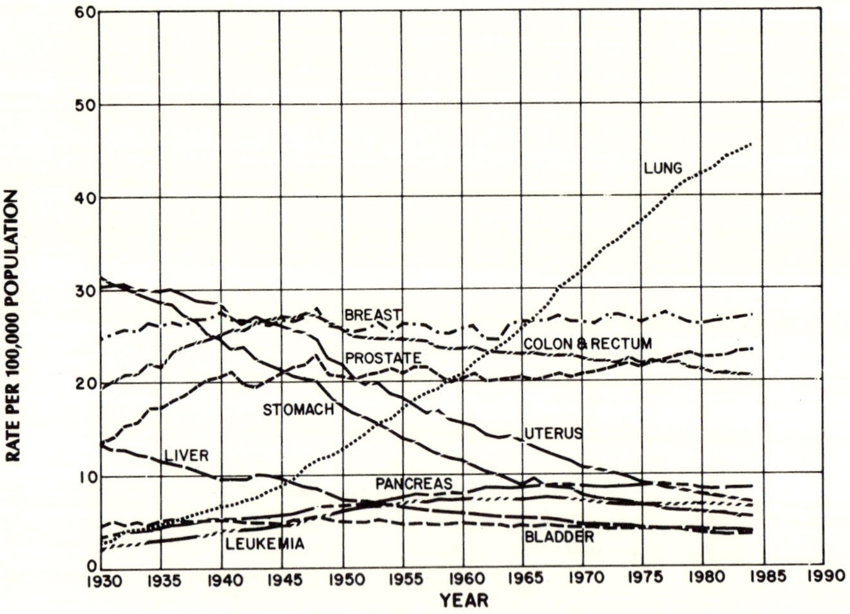

FIG. 8-1. Cancer death rates by anatomical site as a function of year, United States, 1930-1984 [Reprinted, with permission, from "1986 Cancer Facts and Figures" (ACS 1987, p. 11).]

B. Lung cancer among uranium miners

1. Early experience. It had been known since the Middle Ages that in some parts of Southern Germany and Czechoslovakia miners had lung problems, called Bergkrankheit (mountain sickness). The mines were originally for metals other than uranium, but the ground was rich in uranium and at least some of the mining eventually switched to pitchblende, an ore in which uranium is found. The first identification of the lung problem with lung cancer was made in 1879 for the miners at Schneeburg, Germany. From 1869 to 1877, 150 miners there died of lung cancer, out of a work force which numbered about 650 in 1879. Overall, it has been estimated that there were 400 cancer deaths from 1869 to 1935 (Donaldson 1969, Holaday 1969).

Most of these deaths are now associated with the inhalation of radon daughters. However, the role of radon as a causative agent was apparently not suggested until 1924 (Donaldson 1969) and "not generally accepted until the 1960's" (NCRP 1984b, p. 91). While this seems surprising, the occurrence of other lung ailments among miners, such as black lung disease among coal miners, may have made the special role of radon less obvious.

Radon concentrations in the Schneeburg mines have been subsequently estimated to have been 2900 pCi/ℓ in one estimate and 15,000 pCi/ℓ in another (Holaday 1969). Thus, radon levels were very high, although not known precisely. Lung cancer incidence was also high. The recognition that the medical problems encountered at Czechoslovakian mines (in Joachimstal) were the same as those at Schneeburg was not made until 1926 (Holaday 1969). Following this association, autopsy studies in 1929 and 1940 showed that 50% of the miners died of lung cancer. The radon levels were probably similar to those at Schneeburg (Holaday 1969).

Accurate, quantitative correlations between exposure levels and incidence of lung cancer cannot be retrospectively established for these early, pre-World War II, exposures. However, detailed studies of post World War II experience have been undertaken in a number of countries. These are reviewed in Sections B-2 and B-3, below.

2. Studies of U.S. uranium miners. There have been very extensive studies of U.S. uranium miners, starting after the end of World War II when large scale use of uranium began, first for weapons and then for nuclear power. These studies were initiated in 1949 under the auspices of federal and State of Colorado health agencies and, in the words of one of the pioneers in these studies, "with the possible exception of free silica, no other toxic material has had more different groups examining such extensive information" (Holaday 1969).

The first studies, in which radon levels were measured for 34 mines in Utah and Colorado, showed radon concentrations extending as high as 50,000 pCi/ℓ with typical values being several thousand pCi/ℓ. Somewhat lower concentrations were observed in mines on the Navajo Reservation. An awareness of the need for controls followed shortly, and by 1955 a conference was held in Salt Lake City on the health hazards of uranium mining. It was at this meeting that the recommendation was made that concentrations be held below a newly adopted "working level", equivalent to 100 pCi/ℓ of radon in equilibrium with its daughters (Holaday 1969). By 1960, the lung cancer deaths of 9 uranium miners were associated with the exposure to radon and its daughters, and control measures were intensified. In 1961 almost 60% of the mines still had concentrations above 3 WL, with 29% above 10 WL, but by 1967 the fraction above 3 WL had dropped to 6% and only 1% remained above 10 WL (Holaday 1969).

Starting in the late 1960s, a succession of federal agencies has promulgated standards limiting the exposures of miners to approximately 4 working level months (WLM) per year, corresponding to a concentration of about 0.3 WL. The limit of 4 WLM per year became the Environmental Protection Agency standard in 1971, and this standard remains in effect today (Morgan 1986). Actual exposures of miners are estimated now to average about 1 to 2 WLM per year (NCRP 1984b).

Detailed analyses of health effects were begun during the 1950s, initially for a group of 5,370 uranium miners and mill workers (Donaldson 1969). Subsequently the group being studied was reduced to uranium miners alone, including 3,414 white miners and 780 American Indian (mostly Navajo) miners (Archer 1973). This represented approximately 90% of the U. S. uranium miners in the Colorado plateau region as of 1950. For each person studied, an attempt was made to reconstruct his exposure, although this is complicated by the fact that during the period of the exposure to the highest levels, before 1960, the data on radon concentrations were not complete. Further, many of the uranium miners had prior records of work in other mines, where they might have been exposed to radon or other carcinogenic agents.

Some of the main findings as of January 1974 were (Archer 1973, 1974, 1976): (1) for white miners, there were many more lung cancer fatalities than would have been expected without radon exposure (174 observed vs. less than 30 expected); (2) cancers usually occurred more than 10 years after the person started mining, and almost never sooner than 5 years after; (3) there was apparently no significant increase in lung cancer for relatively low exposures (120 WLM and less).

The 780 American Indian miners were treated separately since the lung cancer incidence in this cohort is much less than in the white U.S. population (Samet 1984; Archer 1976). This is primarily attributed to the low incidence of smoking among the Indian population. As a result, the Navajo miners represent a significant population in which to assess the risk of radon-induced lung tumors. In the early studies no significant increase in lung cancer deaths were seen in Navajo miners because of the small population size. However, subsequent studies with longer follow-up times (25 years) have demonstrated a significant excess of lung cancer deaths in this population with 11 observed as compared to an expected number of less than 3. Samet (Samet 1984) has also done case-control studies of 32 Navajo Indians who died from lung cancer and found that 23 had worked as uranium miners with a median lifetime exposure of 1,207 WLM. In addition, these Navajo miners developed lung tumors at a median age of 44 years whereas non-miners developed lung tumors at a median age of 63 years, which is similar to what is seen in the U.S. white population. These data, therefore, confirm a close association of lung cancer with radon daughter exposure.

An analysis of the mortality data through September 1974 was made as part of a comprehensive study of ionizing radiation effects by the Committee on the Biological Effects of Ionizing Radiations of the National Research Council, commonly know as the BEIR III Report. The results of this analysis (NAS 1980) are presented in Table 8-1. The miners were divided into categories, by estimated exposures in WLM. The categories ranged from under 120 WLM to over 3,719 WLM. Under present-day conditions it is unlikely that a miner would receive a lifetime exposure as high as 120 WLM, but in the early years of U.S. uranium mining typical exposures were much higher.

TABLE 8-1. *Lung cancer risk for U.S. uranium miners due to radon exposure.*

(1) WLM Interval	(2) WLM Midpoint	(3) Person Years	(4) Lung cancers Observed	(5) Lung cancers Expected	(6) Lung cancers per WLM per 10^6 PY
0-119	60	5,183	3	3.96	—
120-239	180	3,308	7	2.24	8.0
240-359	300	2,891	9	2.24	7.8
360-599	480	4,171	19	3.33	7.8
600-839	720	3,294	9	2.62	2.7
840-1,799	1,320	6,591	40	5.38	4.0
1,800-3,719	2,760	5,690	49	4.56	2.8
>3,719	7,000 (est.)	1,068	23	0.91	3.0
ALL	1,180	32,196	159	25.24	3.52

Reference: NAS 1980, p. 319.
Notes:

1. Specifies range of cumulative exposures over working lifetime.
2. Mid-point of interval from (1).
3. The number of person-years is based on the number of years at risk. For each person, the years-at-risk is taken to equal the time interval starting 10 years after the beginning of mining and continuing to the end of the study period (September 30, 1974).
4. Observed lung cancers include those during the time interval starting 10 years after the beginning of mining and continuing to the end of the study period (September 30, 1974).
5. Expected lung cancers are based on "age- and year-specific rates for white males in Colorado, Utah, New Mexico, and Arizona."
6. Sample calculation (for 360-599 WLM):
   $(19 - 3.33)/(480 \times 4171) = 7.8 \times 10^{-6}$ per WLM per person-year

The risks are expressed in terms of fatal cancers per million person-years (PY) per WLM. (The meaning of this terminology can be illustrated by an example: If an individual has an exposure of 3 WLM in one year, and the stated risk is 7 per 10^6 person-years per WLM, then during a 30-year subsequent period his risk of dying of lung cancer due to the one-year exposure would be: $3 \times 30 \times (7 \times 10^{-6}) = 6.3 \times 10^{-4} = 0.063\%$. In this illustrative example, issues of latent period and cell repair are ignored; they are considered in the more comprehensive discussion of Chapter 9.) Conspicuous features of the data presented in Table 8-1 are:

1. The highest risk per WLM is for intermediate total exposures (120 WLM to 600 WLM).
2. At higher exposures, the risk per WLM is less although the total risk is higher.

3. At exposures below 120 WLM, there is no evidence of excess cancers among U.S. miners. In fact there was one fewer death than statistically expected, but with the small numbers involved this result is not statistically meaningful, nor can some cancer excess be ruled out.

3. Epidemiological analyses by the NCRP. The results of studies of miners in a number of countries were compiled and analyzed in a report (NCRP 1984b) issued in 1984 by the National Council on Radiation Protection and Measurements. The United States results were essentially the same as those cited above from the BEIR III report. The studies for other countries are cited briefly below:

(a) Uranium miners in Czechoslovakia. The post World War II mortality among these miners has been studied by Czechoslovakian scientists (see, e.g. Kunz 1979). A particular subset of their data was selected for review: those who started mining in the years 1948 through 1952 and who were at least 30 years old at this time. Younger miners are excluded because there was insufficient time for most of the eventual cancers to appear (lung cancer rarely appears before age 40) by the end of 1975, the latest date for which lung cancers had been surveyed. About 2400 persons were in this group, with an average observation period of 23.5 years (including those under 30).

(b) Swedish miners. Studies were made of radon concentrations in Swedish iron, zinc and lead mines during 1969 and 1970. Radon concentrations were found as high as 20,000 pCi/ℓ, although the average was under 100 pCi/ℓ.

(c) Fluorspar miners in Newfoundland. High radon levels were encountered in fluorspar mines in Newfoundland. The radon was brought in by ground water, rich in radon.

(d) Canadian uranium miners (Ontario). Perhaps because uranium mining did not start in Canada until 1954, close to the time that sensitivity was rising as to the dangers of radon, the individual exposures were less for the Canadian miners than for the U.S. miners.

The results of these studies, along with those from more recent studies of Swedish iron miners (Radford 1984) and of Saskatechewan uranium miners (Howe 1986) are summarized in Table 8-2. Much of this body of data is also presented in graphical form in Figure 8-2, where the risk is displayed as a function of total exposure. Several points stand out in the consideration of these data:

1. For all groups, high radon exposures lead to an unambiguous excess of fatal lung cancers over the number of lung cancer deaths expected for an unexposed population.
2. At a given exposure level (measured in total WLM), there is a wide disparity in the consequences reported in different studies, with the United States studies showing the lowest cancer incidence per WLM.
3. In each of the studies considered in NCRP Report No. 78, the cancer incidence per WLM is least at the lowest total exposures.

TABLE 8-2. *Estimates of lung cancer mortality among miners due to inhalation of radon daughters.*

Group	Lung cancers/WLM-10^6 PY		
	NCRP	Radford	Howe
U.S. uranium miners	0-8	6-9	—
Czech uranium miners	5-23	18-21	—
Swedish iron miners, Malmberget	19	19	—
Swedish miners, other	7-35	—	—
Canadian uranium miners	4-11	6-15	—
Newfoundland fluorspar miners	18	18	—
Canadian uranium miners, Saskatchewan	—	—	21

Notes:

1. The data in Column (2) are extracted from NCRP 1984b, Table 8.1. Where results are presented for different exposure bands, the tabulated range of results corresponds to the maximum and minimum values found for the various bands (with negative values changed to zero).
2. The data in Column (3) are from Radford 1984, Table 6, based on a linear fit to the results for exposures of less than 600 WLM.
3. The data in column (4) are from Howe 1986, Table 4. The single tabulated value corresponds to the average over bands extending from 5-24 WLM to over 250 WLM.

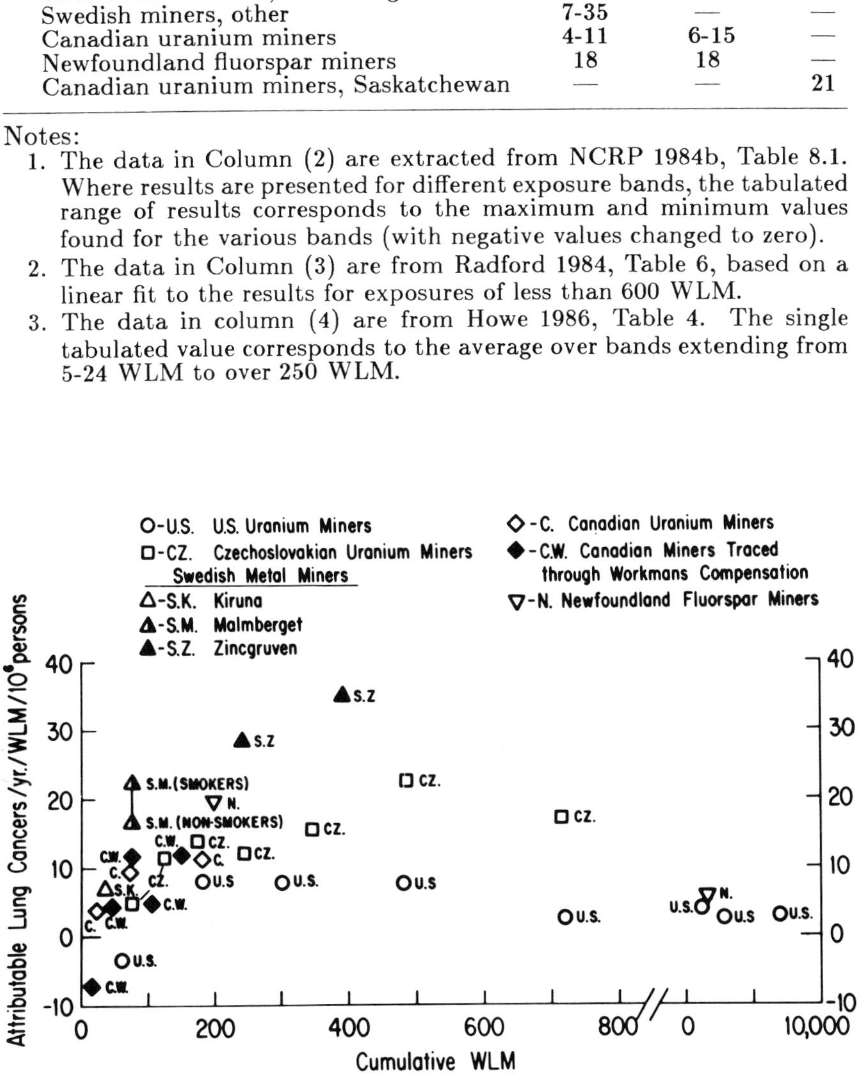

FIG. 8-2. Lung cancer risk per WLM as a function of cumulative exposure. The risk is expressed as the attributable annual risk per WLM per million persons. [Reprinted, with permission, from NCRP 1984b, Figure 8.2.]

In contrast, recent studies of Swedish iron miners (Radford 1984) and of Canadian uranium miners (Howe 1986) do not show a falloff in this rate at low exposure levels.

4. *Cancer incidence at low exposure levels.* The epidemiological data from studies of miners do not resolve the question of paramount interest in the context of indoor radon—the rate of cancer incidence for low exposure levels. Only rarely will indoor radon levels exceed several WLM per year (the national average is roughly 0.2 or 0.3 WLM per year) and the typical lifetime exposure is usually less than 20 WLM.

A compilation of the data for lifetime doses below 120 WLM is presented in Table 8-3. It is seen that the United States and one of the Ontario studies show a deficit of lung cancers at the lowest exposures while the Czechoslovakian, Swedish and Saskatchewan studies show an excess. The numbers involved are small and it is not possible to draw any definitive conclusion.

At the time of the NCRP review of the data, the analyses from one of the Swedish studies (study "B") and the Saskatechewan study were not available. These studies show the greatest lung cancer excesses at low dose rates. Without these data, and even with them, it might be argued that there is a threshold below which radon exposure does not lead to cancer. However, the NCRP Report concludes that (NCRP 1984b, p. 111):

> ... [the available data] are not suitable to determine whether a threshold exposure for lung cancer induction exists. In order to estimate lung cancer risk in exposed populations it is in keeping with present day views of radiation biology and radiation protection to assume that radiation induced-cancer is a stochastic process correlated with dose and without a threshold.

The Report goes on to the additional conclusion for the levels typically experienced by the general population [ibid, p. 112]:

> ...in estimating the effect of radon daughter exposure at environmental levels, normally less than about 20 WLM per lifetime, the attributable risk at high exposure levels must be extrapolated to the low exposure region.

In these sentences, the NCRP Report argues against assuming a threshold and indicates that effects at low dose levels are to be calculated by extrapolating from observed effects at high dose levels. At a later point (see Chapter 9), the Report adopts a specific extrapolation based upon the so-called linearity hypothesis. The meaning of this hypothesis is discussed in the following Section.

TABLE 8-3. *Lung cancer risk due to radon for lifetime exposures of less than 120 WLM*

(1) Group	(2) WLM category	(3) Lung cancers observed	(4) Lung cancers expected	(5) Risk per WLM per 10^6 PY
Saskatchewan[3]	12	12	6.5	32
Ontario uran. B[1]	15	38	56	-7
Ontario uran. A[1]	21	40	33	3.8
Swedish iron B[2]	27	8	3.4	19.3
Swedish iron A[1]	34	13	4.5	7
Saskatchewan[3]	36	5	2.6	12
Ontario uran. B[1]	46	23	14	4
U.S. uranium[1]	60	3	3.96	-3.1
Saskatchewan[3]	70	6	2.5	13
Ontario uran. A[1]	72	16	7	9.6
Czech uranium[1]	72	6	3.1	4.6
Swedish iron B[2]	73	14	2.6	20.3

References: [1]NCRP 1984b, Tables 8.1 and 8.4; [2]Radford 1984, Table 4; [3]Howe 1986.

Explanation of columns:

1. The data for U.S. miners are for white miners only. Two different studies are presented for Ontario uranium miners (A and B) and for Swedish iron miners (A and B).
2. The quoted midpoint in column (2) represents, variously, the mean or median of the band studied.
3. For most of the studies, the lung cancer data is based primarily on cancer mortality statistics, although the U.S. results include some still-living lung cancers cases. For the Ontario B study, the data refer to lung cancer incidence.
4. The rate of lung cancers per WLM per person-year is calculated from the number of excess cancers, using the total number of years at risk. In calculating the number of person-years, somewhat different assumptions about latent periods have been made in the different studies.

C. Dose-response effects and the linearity hypothesis

1. The linearity hypothesis at low dose levels. The effects of radiation exposures at high dose levels have been established from direct observational evidence, especially in studies of victims of Hiroshima and Nagasaki, of accident victims, and of individuals who received high doses during medical irradiations. Whole-body doses above about 400 rem, received over a short period of time, are likely to cause early death from acute effects. If the doses are somewhat reduced or spread over a period of time, the main impact is that of cancer induction. For example, the increased lifetime risk of fatal cancer is estimated to be several percent for a dose of 100 rad of sparsely ionizing radiation (equivalent to 100 rem) (NAS 1980).

There is no direct evidence bearing on radiation-induced cancer mortality for individuals receiving doses which are, say, 100 times lower. To determine the cancer induction rates at low dose levels it is necessary to extrapolate from the information established at high doses. The linearity hypothesis provides one means of making the extrapolation. In this hypothesis, it is assumed that the magnitude of the effect of interest, in this case the number of cancers produced, is directly proportional to the dose received. Thus, for example, comparing effects at high and low radiation doses, at one-hundredth the dose the chance of cancer incidence will be one-hundredth as great.

The meaning of linearity can be understood in terms of graphs of cancer incidence vs dose, as in Figure 8-3. The straight line, curve (2), is a graph of a linear response. It is characterized by the expression:

$$\text{cancer incidence} = \text{constant} \times \text{dose}.$$

A slightly more general dose response curve has the form:

$$\text{cancer incidence} = \text{constant} \times (\text{dose})^a.$$

For linearity, $a = 1$. Curve (1) of Figure 8-3, on the other hand, is drawn with $a = 0.8$ and curve (4) with $a = 2$.

The extrapolation from high doses to low doses can be considered in the context of Figure 8-3. Suppose that the cancer incidence rate is measured at a dose of 100 rad. This establishes one known point. A second point is at the origin, because if the dose is zero the cancer rate is zero (taking both dose and cancer rate to refer to increments over normal conditions). These two points can be connected by a variety of theoretical curves, e.g. curves (1) to (4). These curves imply very different results for the cancer incidence at intermediate doses. Thus, for example, the cancer incidence at 20 rad is much greater for curve (1) than for curve (4). Curve (2), corresponding to linearity, lies between.

There is considerable disagreement as to the true form of the dose response curve. In the BEIR III Report (NAS 1980), a compromise consensus was reached in which linearity was adopted for evaluating cancer incidence due to densely ionizing radiation such as alpha

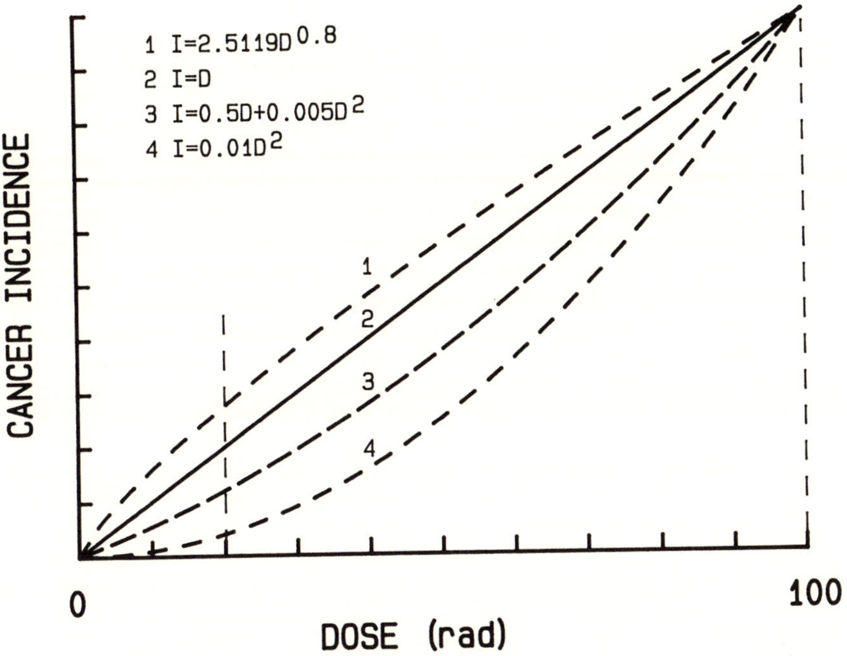

FIG. 8-3. Illustrations of alternative dose-response models. Curve (2) corresponds to the linearity hypothesis, while the other curves illustrate examples of the so-called supralinear (curve 1), linear-quadratic (curve 3), and quadratic (curve 4) models. The curves are constrained to the same incidence of cancer at a dose of 100 rad.

particles, while for sparsely ionizing radiation, in particular for beta particles and gamma rays, an expression was favored corresponding in general form to curve (3) of Figure 8-3.

Despite the adoption of the linearity assumption for alpha particles, it is suspected among some radiation scientists that even in this case the linearity assumption overestimates the lung cancer incidence from radon at low doses. But this conclusion is by no means well established or universally accepted, as illustrated by the unwillingness of the NCRP panel to discard the linearity hypothesis despite hints in that direction from the studies of cancer incidence among miners. Similarly, even for sparsely ionizing radiation, the BEIR III Report included consideration of a linear response, as a plausible, if probably extreme, possibility.

Although a linear extrapolation from the miner epidemiological data to the much lower radon daughter exposures received by most of the general population may overestimate the lung cancer risk, it also is possible that it may underestimate the risk. Some animal experiments with neutrons, which like alpha particles lead to high ionization densities, show a downward bending of the dose-life shortening response curve similar to curve (1) of Figure 8-3 (Fry 1981). Nearly all radiation induced life shortening in experimental animals at moderate dose levels and in the Japanese survivors is the result of deaths from cancer, and thus life shortening studies are useful guides to interpretation of the shapes of dose-cancer response curves. In radon daughter exposed rats, down to 65 WLM, the dose-response for lung tumors was found to be consistent with linearity in dose or possibly with a downward bending of the curve [i.e., dose exponent less than 1] (Chmelevsky 1982). More recent U.S. (Cross 1986) and French (Chmelevsky 1984) studies down to 20 WLM suggest a linear response below 50 WLM.

In summary, what direct evidence there is on the incidence of radon-induced cancer at low dose levels comes from studies of animals not humans. In the interests of caution, the NCRP and most other authorities adopt a linear extrapolation from the observed rates at higher exposure levels. This represents a prudent approach, especially in terms of setting standards for population protection. Nevertheless, when this linear extrapolation is used for estimating deaths from indoor radon, typically at exposure levels of under 1 WLM per year, it should be recognized that the underlying assumption of linear response is not based on firm empirical evidence in humans.

2. Effects at high dose levels. Miner epidemiological studies suggest that lung cancer incidence increases with increasing radon daughter exposure, reaches a maximum, and then decreases with still higher lung exposures as has been seen in Figure 8-2 for miners. Combined results for humans and experimental animals are presented in Figure 8-4. This effect of decreasing cancer incidence with increasing large doses of radiation is not unique to lung cancer. It has been reported for other types of radiation carcinogenesis in animals (Hellman 1982) and in cells grown outside the body (Borek 1973). This may be due, at least in part, to greater and selective cell killing by large radiation exposures (Borek 1973; Leenhouts 1978). Selective cell killing implies that those cells which have suffered injury which makes them cancerous also have sustained lethal cellular injury. A dead cell cannot multiply and develop into a tumor.

It should, of course, be noted that the decreasing cancer rate is not seen until cumulative exposures exceed at least 500 WLM, and it is not anticipated that such exposures will be reached for indoor radon, except under very unusual circumstances.

D. Other biological factors affecting radon-induced lung cancer

1. Occurrence of lung cancers by type. In the classification of lung cancers by cell types, lung cancers are divided into four general categories: (1) epidermoid or squamous cell carcinoma (platelike cells similar to those of the skin), (2) small cell carcinoma, (3) adenocarcinoma (a cancer having a glandular-like structure), and (4) large cell

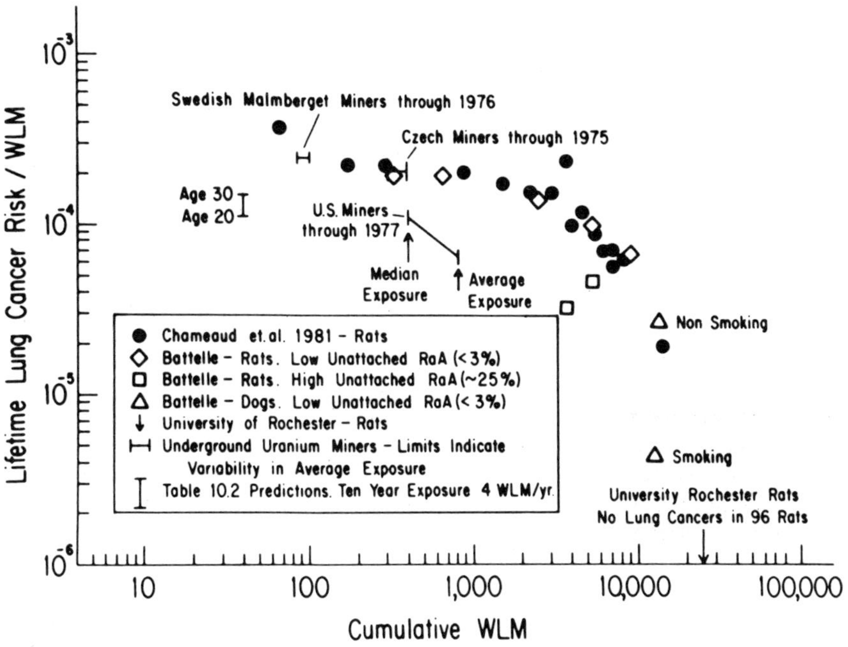

FIG. 8-4. Lifetime lung cancer risk per WLM in humans and animals as a function of cumulative exposure. [Reprinted, with permission, from NCRP 1984b, Figure 11.1.]

carcinoma. For lung cancers in the United States, as determined at autopsy, the highest percentage is epidermoid carcinoma (about 1/3) and the lowest percentage of these is large cell carcinoma (about 1/6) (Straus 1977).

Lung tumors occur in various regions of the lung, but most are associated with the bronchi (tubes that carry air deep into the lungs) (Seydel 1975). It has been estimated that 1/3 of the epidermoid carcinomas are deep in the lung tissue and not located in the main air passages. In contrast, those cancers arising in uranium miners do not usually involve the deep lung tissue, but rather sites of the cancers tend to be located in certain of the main air passages as discussed in Chapter 7.

As pointed out in the NCRP Report No. 78: "It would simplify the epidemiology if the lung cancers induced by radon daughters were different from those occurring spontaneously or those induced by smoking" (NCRP 1984b, p. 93). Early studies suggested a predominance of cancers of the small cell undifferentiated type. However, this effect has not appeared consistently in subsequent studies, and there are indications that the proportion of different histological types depends importantly on the length of the latent period and on the smoking history. In the absence of detailed individual information on these factors, the Report concludes that "cell type is not a useful method of differentiating cancers attributable to radon daughters from those occurring in the general population" (NCRP 1984b, p. 95). Overall, it does not appear that the distribution of cancer cell types is markedly different following radon exposure as compared to lung cancers from other causes, although as indicated in the preceeding paragraph, the distribution by anatomical site differs to some extent.

2. Synergistic effects of smoking. A complicating factor in interpreting radon epidemiological data is the contribution from smoking. Account should be taken of the smoking patterns, but these are not always well known in the populations being studied. In addition, it is not known if the combined effects of radon and smoking are worse than the sum of the individual effects, or, in other terms, if a given exposure to radon causes more additional cancers among smokers or among non-smokers. If smokers are at greater incremental risk from a given radon exposure than are non-smokers, a synergistic effect is said to exist between smoking and radon exposure. In calculating total risks, in the absence of synergism the risks from smoking and radon are additive. With synergism, the effects are more complicated but they could be, for example, multiplicative.

On the simplest level, it is clear that high levels of radon exposure can cause lung cancer, independent of smoking. Studies of Navajo miners, who predominantly are non-smokers, have shown a significant excess of lung cancer, presumably due to radon daughter exposure (Samet 1984). However, for more complex combinations of smoking and radon, the risk evidence is contradictory.

For example, an analysis of the experience of U.S. uranium miners concluded that the results "provide strong and consistent support for a description of lung cancer risk as the product of components due to radiation and cigarette smoke" (Whittemore 1983). Similary, synergism was favored in a study of Swedish iron miners which concluded that "most of the data...indicates an enhancing effect of smoking, approximately of a multiplicative type" (Damber 1982). But, in contradiction, two other studies of Swedish iron miners (Edling 1983, Radford 1984) conclude that the impacts of smoking and of radon are additive, or nearly so, i.e. they act independently with no significant synergistic effect. At the extreme, it has been suggested, on the basis of a study of Swedish zinc-lead miners, that smoking may reduce the effect of radon, i.e. have a protective action (Axelson 1978). This might be understood if smoking led to a thicker mucus layer over the bronchial surfaces, reducing the dose caused by the alpha-particles emitted by radon daughters lodged on these surfaces.

Animal experiments do not resolve the issue. A study of beagle dogs (NCRP 1984b) also has suggested a protective effective of smoking. However, a positive synergistic effect was found in experiments in which rats were exposed to diferent radon levels, with and without cigarette smoke added (Chameaud 1982).

Overall, recent reviewers of the array of evidence have tended to favor, albeit tentatively, some degree of synergism. Thus, study of the epidemiological data for uranium and other miners has led one group of authors to suggest that "the interaction between radiation and smoking may be intermediate between additive and multiplicative" (Thomas 1985) and another group to conclude that the evidence favors a synergistic model but that "present data do not allow emphatic acceptance of any model" (Ginevan 1986). Still a third review reaches no firm conclusions and cautions that the "data from the various [miners] populations are not consistent" (Samet 1986), but tends to place somewhat more credence in a multiplicative model.

The difficulty facing any analysis of possible synergism is that the available study populations are small and, perhaps in part for that reason, the data for different individual populations are not mutually consistent. It will not be easy to improve the data base, because radon levels in mines have been reduced over the past several decades and therefore analyses of more recent experience will have even less statistical accuracy. Of course, the resolution of the issue would have important implications for determining the relative signficance of radon to smokers and non-smokers. In particular, if a strong enough synergism were to be established, the "radon problem" would become largely a "smoker's problem."

The issue of synergism is closely related to the distinction between "absolute risks", corresponding to additive risks (i.e., no synergistic effects) and "relative risks," corresponding to multiplicative risks (i.e., synergistic effects). Absolute and relative risks will be discussed further in Chapter 9, including consideration of the differences in the implications of the two models for smokers and non-smokers.

3. Latent period before the onset of cancer. The studies of miners have also shown that lung cancer incidence from radon, as with cancers from other sources of ionizing radiation, is delayed some years after the exposure. As summarized in NCRP Report No. 78 (1984b, p. 111):

> The latent period seems to vary inversely with age at first exposure, with amount of cigarette smoking, and with total exposure and/or exposure rate. That is, the shortest latent periods are found among those men who are elderly at start of mining, who smoke heavily, and who have the most intense exposures. The latent period has a large range of about 7 to 50 years. Mean values are usually considered to be between 20 and 30 years, but one study [of Swedish miners] reported a mean of 43 years among non-smokers exposed at low levels who had been followed for over 60 years. In none of the studies so far has there been any significant appearance of lung cancers before age 40.

Some, but not all, of these complexities in the latent period are incorporated in the NCRP predictive risk model, discussed in Chapter 9.

E. Epidemiological studies of the general population

1. Some difficulties in epidemiological studies. In principle, the impact of radon exposures on the general population can be investigated directly by comparing lung cancer incidence in areas with high and low radon concentrations. Analogous studies for other sources of natural radiation exposure have not convincingly established correlations between cancer and radiation level, probably because natural radiation levels are generally too low. However, typical doses to the lung due to radon are much greater than the typical dose to any organ from other natural sources of radiation, and therefore correlations may show up more strongly in the case of lung cancer from radon.* In addition, there are large variations in radon exposure levels, facilitating comparisons.

However, despite the promising potential for correlation studies, as yet there are no well established conclusions. The available data on radon levels are still too fragmentary and there are major confounding problems. For one, smoking is a considerably more important cause of lung cancer than radon. Even neglecting variations in smoking habits, the presence of a high "background" due to smoking would lessen the statistical accuracy of analyses of radon effects and, of course, the smoking variations themselves can be crucial.

There are other problems as well. The cancer data used to date have been total lung cancer statistics for a given area, for the United States usually a county, with no examination of the details of the exposure of particular individuals to radon or to other possible causes of lung cancer. The impact of radon depends upon the accumulated exposure over a lifetime, and even were the average present radon level in a given region well known, this alone does not establish the lifetime exposure history of the individuals who die of cancer in that region. Individuals move from region to region and for those individuals who have stayed within the same region exposures may have varied due to changes in house construction and tightness.

2. Review of existing studies. Nevertheless, despite these and other difficulties, it is of interest to consider the status of the existing epidemiological evidence. This evidence has been recently reviewed by Cohen (1987) and by Hofmann *et al.* (1986). We summarize below some of the pertinent studies, including their own:

---

* Typical lung doses from radon are in the neighborhood of 2000 or 3000 mrem per year, while typical doses from other all natural sources of radiation are about 100 mrem per year (see Chapter 10).

(a) An extensive program to explore correlations between lung cancer and radon has been undertaken by Cohen (1987) at the University of Pittsburgh. To minimize the confounding effects of smoking, these studies concentrate on observed lung cancer rates among females during the period 1950-69. During this period, the female lung cancer rate was much below that for males, presumably because women smoked less. Cohen's preliminary results show no positive correlation between high radon levels and lung cancer incidence. In fact, for Cumberland County, Pennsylvania, where the average radon level is estimated to be about 7.5 pCi/ℓ (about 5 times the national average), the lung cancer rate during 1950-69 was below the national average (5.4 cancers per year per 100,000 women in Cumberland County vs. 6.2 nationally). In a separate study, which included localities in all parts of the country, the 10 localities (out of 101) which had the highest radon concentrations had below average lung cancer rates.

(b) Lung cancer incidence has been studied for two areas of Guangdong Province in China (Hofmann 1986). The total alpha-particle exposures (including contributions from thoron (radon-220) and outdoor radon, which were found to be appreciable) were 0.38 WLM per year for the "high-background area" and 0.16 WLM for the "control area". During the period from 1970 to 1983, there were 23 lung cancer deaths in the "high-background" area and 27 in the "control" area, for virtually the same number of person-years of observation. For each group, both with more men than women, the annual lung cancer mortality rate was under 3 per 100,000 persons, which is far less than for most countries (see Section A) and, for the "high-background" area, is less than one-fifth of the rate predicted for radon alone using a linear extrapolation from the miners' experience.

(c) An area of Southeastern Finland has been studied, for which the present mean indoor radon concentration (10 pCi/ℓ) is about 4 times the Finnish national average (Castren 1985). During 1955-74, the average annual lung cancer incidence rate in this area was 72 per 100,000 for men and 5.6 per 100,000 for women, compared to national averages of 82 and 5.2 for men and women, respectively. However, the authors caution that "far reaching conclusions should not be drawn" from the lack of significant dependence of lung cancer rates on present radon levels. They point out that past exposures may have been very different, due to recent additions of weather stripping and central heating, as well as to population movements.

Cohen and Hofmann *et al.* also cite studies for Sweden and Austria which show below average lung cancer rates in regions of high radon concentration. On the other side of the ledger, a study carried out on the Swedish island of Oeland showed a positive correlation between radon level and lung cancer mortality rates (Edling 1984). This study had the advantage of focusing on a rural population with low lung cancer rates and of being restricted to individuals who had lived at least 30 years in the same house, with both the radon level and smoking habits known. However, with these restrictions, the number of lung cancer deaths in the remaining sample was only 19 individuals, split between smokers and non-smokers. Cohen also cites a study in Great Britain where a positive correlation was reported between lung cancer and radon level. A positive correlation is also suggested in preliminary studies of the high-radon Reading Prong region of Pennsylvania by Fleischer (1986), although Cohen reports below average lung cancer rates for the Reading Prong area. A positive, but not statistically significant, correlation between lung cancer incidence and radioactivity in the ground was found in a comparison for regions in Viterbo Province, Italy (Forastiere 1985). A positive effect is also reported by Hess and collaborators (1983) who find higher overall cancer rates and higher lung cancer rates in counties where the average radon concentration in water is high, and where one might therefore expect that radon concentrations in the air are also high.

3. Summary. These studies do not purport to be definitive and at this early stage it is not surprising that the results are not mutually consistent. They suggest the following interim comments:

1. At the radon levels typically encountered by the general public (as distinct from miners), there is no convincing observational evidence that elevated radon concentrations cause an increased number of lung cancers. The little positive evidence advanced for such an effect appears to be at least balanced and possibly outweighed by studies which show no positive correlation.

2. It should be possible to obtain better knowledge as to correlations or lack thereof, given continued studies.

3. Such studies, especially with refinements in methodology and use of larger data bases, offer the hope of giving a quantitative insight into the relationship between radon level and cancer incidence at low radon levels. In fact, the studies of Cohen are directed towards testing the linearity hypothesis, discussed in Section C. His preliminary conclusion, as well as the suggestion from the Guandong Province study, is that the cancer induction rate due to radon is less than would be predicted by linear extrapolation from the experience of miners.

In summary, it is not as yet possible to use reported epidemiological studies of the general population to determine lung cancer incidence due to radon, although there is hope for the future. For the moment, it is generally deemed prudent to rely on estimates based upon extrapolations from the effects found for miners. The implementation of such estimates is discussed in Chapter 9.

References: Chapter 8

American Cancer Society, 1986, *1986 Cancer Facts and Figures* (New York: ACS).

______, 1987, *Cancer Facts and Figures-1987* (New York: ACS).

Archer V. E., Wagoner J. K. and Lundin F. E., 1973, "Lung cancer among uranium miners in the United States," *Health Phys.* **25**, 351.

Archer V. E., Saccomanno G. and Jones J. H., 1974, "Frequency of different histologic types of bronchogenic carcinoma as related to radiation exposure," *Cancer* **34**, 2056.

Archer V. E., Gillam J. D. and Wagoner J. K., 1976, "Respiratory disease mortality among uranium miners," *Annals NY Acad. Sci.* **271**, 280.

Borek C. and Hall E. J., 1973, "Transformation of mammalian cells in vitro by low doses of x-rays," *Nature* **243**, 450.

Castren O., Voutilainen A., Winqvist K. and Makelainen, 1985, "Studies of High Indoor Radon Areas in Finland," *Science of the Total Environment* **45**, 311.

Chameaud J., Masse R., Morin M. and Lafuma J., 1984, "Lung Cancer Induction by Radon Daughters in Rats: Present State of the Data on Low-Dose Exposures," in Proceedings of the International Conference on Occupational Radiation Safety in Mining, Vol 1. (edited by H. Stocker), pp. 350-353 (Toronto: Canadian Nuclear Assn.).

Chameaud J., Perraud R., Chretien J., Masse R. and Lafuma J., 1982, "Lung carcinogenesis during in vivo cigarette smoking and radon daughter exposure in rats," *Recent Results in Cancer Research* **82**, 11.

Chmelevsky D., Kellerer A. M., Lafuma J. and Chameaud J., 1982, "Maximum likelihood estimation of the prevalence of non-lethal neoplasms – an application to radon-daughter inhalation studies," *Radiat. Res.* **91**, 589.

Cohen B. L., 1987, "Tests of the linear-no threshold dose-response relationship for high-LET radiation," *Health Phys.* **52**, 629.

Cross F. T., 1986 (personal communication).

Damber L. and Larsson L.-G., 1982, "Combined effects of mining and smoking in the causation of lung carcinoma," *Acta Radiologica Oncology* **21**, 305.

Donaldson A. W., 1969, "The epidemiology of lung cancer among uranium miners," *Health Phys.* **16**, 563.

Edling C. and Axelson O., 1983, "Quantitative aspects of radon daughter exposure and lung cancer in underground miners," *British J. of Industrial Medicine* **40**, 182.

Edling C., Kling H. and Axelson O., 1984, "Radon in homes—a possible cause of lung cancer," *Scand. J. Work Environ. Health* **10**, 25.

Fleischer R. L., 1986, "A possible association between lung cancer and a geological outcrop," *Health Phys.* **50**, 823.
Forastiere F., Valesini S., Arca M., Magliola M. E., Michelozzi P. and Tasco C., 1985, "Lung Cancer and Natural Radiation in an Italian Province," *Science of the Total Environment* **45**, 519.
Fry R. J. M., 1981, "Experimental radiation carcinogenesis; what have we learned?" *Radiat. Res.* **87**, 224.
Ginevan M. E. and Mills W. A., 1986, "Assessing the risks of Rn exposure: the influence of cigarette smoking," *Health Phys.* **51**, 163.
Hellman S., Moloney W. C. and Meissner W. A., 1982, "Paradoxical effect of radiation on tumor incidence in the rat: implications for radiation therapy," *Cancer Res.* **42**, 433.
Hess C. T., Weiffenbach C. V. and Norton S. A., 1983, "Environmental radon and cancer correlations in Maine," *Health Phys.* **45**, 339.
Hofmann W., Katz R. and Zhang Chunxiang, 1986, "Lung cancer risk at low doses of α particles," *Health Phys.* **51**, 457.
Holaday D. A., 1969, "History of the exposure of miners to radon," *Health Phys.* **16**, 547.
Howe G. R., Nair R. C., Newcombe H. B., Miller, A. B. and Abbott J. D., 1986, "Lung cancer mortality (1950-1980) in relation to radon daughter exposure in a cohort of workers at the Eldorado Beaverlodge uranium mine," *J. National Cancer Inst.* **77**, 357.
Jacob S. W., Francone C. A. and Lassow W. J., 1978, *Structure and Function in Man* (Philadelphia: W. B. Saunders Company).
Kunz E., Sevc J., Placek V. and Horacek J., 1979, "Lung cancer in man in relation to different time distribution of radiation exposure," *Health Phys.* **36**, 699.
Leenhouts H. P. and Chadwick K. H., 1978, "An analysis of radiation-induced malignancy based on somatic mutation," *Int. J. Rad. Biol.* **33**, 357.
Morgan M. V. and Samet J. M., 1986, "Radon daughter exposures of New Mexico U miners, 1967-1982," *Health Phys.* **50**, 656.
National Academy of Sciences/National Research Council, 1980, *The Effects on Populations of Exposure to Low Levels of Ionizing Radiation: 1980,* Report of the Committee on the Biological Effects Press).
NCRP (National Council on Radiation Protection and Measurements), 1984, "Evaluation of occupational and environmental exposures to radon and radon daughters in the United States," *NCRP Report No. 78* (Bethesda, MD: NCRP).
Radford E. P. and Renard K. G. C., 1984, "Lung cancer in Swedish iron miners exposed to low doses of radon daughters," *New England J. Med* **310**, 1485.
Samet J. M., Kutvirt D. M., Waxweiler R. J. and Key C. R., 1984, "Uranium mining and lung cancer in Navajo men," *New England J. Med.* **310**, 1481.
Seydel H. G., Chait A. and Gemelich J. T., 1975, *Cancer of the Lung*

(New York: John Wiley).

Straus M. J., editor, 1977, *Lung cancer, clinical diagnosis and treatment* (New York: Grune and Stratton).

Thomas D. C., McNeill K. G. and Dougherty C., 1985, "Estimates of lifetime lung cancer risks resulting from Rn progeny exposure," *Health Phys.* **49**, 825.

Whittemore A. S. and McMillan A., 1983, "Lung cancer mortality among U.S. uranium miners: a reappraisal," *J. National Cancer Inst.* **71**, 489.

Chapter 9

CALCULATED LUNG CANCER MORTALITY DUE TO RADON

David Bodansky, Kenneth L. Jackson and Joseph P. Geraci

A. Introduction

1. Extrapolation from the experience of miners . From the studies of miners, discussed in Chapter 8, it is clear that an exposure to high concentrations of radon daughters leads to an increased incidence of lung cancer. In this Chapter we will discuss how the experience for miners is used to estimate the incidence of radon-induced lung cancer among the general population.

Such estimates cannot be very precise, in part because the miners studies do not lead to a well-determined relationship between the rate of cancer induction and the exposure levels. In particular, there are disparities in the results of studies for different groups of miners (see Table 8-2). These disparities may arise from the difficulties in obtaining accurate crucial data (NCRP 1984b), such as the year-by-year exposure history over the working lifetime of the individual miners. Other differences may arise from differences in the smoking patterns in different groups and in details of the handling of the data. In any event, there remain substantial uncertainties in the conclusions for the miners themselves, both as to the "average" rate of cancer induction per WLM and as to the variation of this rate with the magnitude of the total exposure.

Further, there are problems in using results obtained for miners at relatively high radon exposure levels for the prediction of the effects of radon at the much lower levels encountered in the normal indoor environment. Here one faces the problem of extrapolating from large dose levels to small dose levels. The generally adopted solution is to assume linearity although, as discussed in Chapter 8, this assumption may overestimate (or, in principle, underestimate) the true risk.

2. Terminology: lung cancer incidence, premature deaths, and life shortening . The effects of radon are expressed in the literature in terms of both "lung cancer incidence" and "lung cancer deaths". Numerically, the two are almost equal, because 87% of lung cancer patients die within five years of diagnosis (ACS 1987). In consequence, it is not always clear if an author is preserving the distinction or if the terms are being used interchangably (see, e.g. NCRP 1984b). This has no significant implications for predicting the effects of radon, given the large uncertainties in predicting either incidence or deaths.

In common usage, the phrase "deaths due to radon" is intended to mean "premature deaths due to radon". A more quantitative concept is that of life-shortening. An analysis of the epidemiology of radon-induced lung cancer (Ginevan 1986) suggests that on average a premature lung cancer death corresponds to a life-shortening in the neighborhood of 15 to 20 years. In the remainder of this chapter, the discussion will be couched in terms of "deaths", which may be a more familiar concept than life-shortening and which is used in most analyses of radon impact.

B. Lung cancer risk calculated by the NCRP

We here review the calculations made in a report prepared by a Task Group of the National Council for Radiation Protection and Measurements, published as NCRP Report No. 78 (1984b). The calculation is based on a model developed by Harley and Pasternak (1981), with some changes in parameters. It is discussed in some detail here to illustrate issues which must be addressed in any estimate of radon consequences.

The calculation takes as a starting point a cancer incidence rate of 10×10^{-6} cancers per WLM per person-year for miners, a rough average of the rates presented in Table 8-2 and Figure 8-2. The following further assumptions are also made:

1. It is assumed that no lung cancers occur before age 40.
2. It is assumed that no radiation-induced lung cancers occur within 5-years of exposure, i.e. that there is a 5-year latent period.
3. Repair mechanisms are assumed, such that the lung cancer probability from a given exposure decreases exponentially with time (after the 5-year latent period), with a 20-year half-time.
4. The contributions to lifetime risk are ignored after the person reaches age 85.
5. It is assumed that the risk per WLM is 40% greater for the general population than for miners, due to the combined effects of differences in average breathing rates, average lung sizes, aerosol particle sizes, and in the fraction of unattached radon daughters (greater in the general environment than in mines due to lower dust levels). It is estimated, overall, that the average dose for the indoor environment is 0.7 rad per WLM, compared to 0.5 rad per WLM for miners.

The exponential factor, contained in assumption (3), corresponds to the assumption of a "decrease in rate of risk expression due to repair, cell death, or unspecified mechanisms", with a 20-year half-time (NCRP 1984b, p. 153). With this choice of half-time, the risk estimates match the results of the studies of miners, where it is found that the incidence of lung tumors is higher when the first exposure to radon daughters occurs at 40 or 50 years of age, as compared to first exposure at age 20 (NCRP 1984b, p. 151). Since lung tumors do not usually appear before age 40, exposure at the earlier age allows more time for decay of the risk. However, the model could be equivalently adjusted to match the observations by assuming that the vulnerability is greater if exposure occurs at a later age (Harley 1981).

With these assumptions, the probability of cancer induction for a person between the ages of 40 and 85 is taken to be:

$$A(t, t_o) = 14 \times 10^{-6} \times N(t_o) \times P(t, t_o) \times e^{-T\ell n2/20} ,$$

where:

$A(t,t_o)$ = the probability of cancer induction at age t, due to an exposure to N WLM occurring at age t_o.
$N(t_o)$ = the number of WLM of exposure at age t_o, where exposure is taken to be in a typical indoor atmosphere.
$P(t,t_o)$ = the probability that a person alive at age t_o will still be alive at age t.
$T = t - t_o$ = the time interval since the exposure.

For example, for a person exposed to 4 WLM at age 30, the calculated risk of contracting lung cancer during the person's 40th year is:

$$A(40, 30) = 14 \times 10^{-6} \times 4 \times 0.98 \times 0.71 = 3.9 \times 10^{-5} ,$$

setting $N = 4$, $t_o = 30$, and $t = 40$. The factor 0.98 represents the fact that about 2% of people alive at age 30 die (of all causes) by age 40 and the factor of 0.71 expresses the effect of repair over 10 years.

The lifetime effect of this single dose at age 30 is found by summing the terms $A(t, t_o)$ for the period from age 40 through age 85 ($t = 40$ to $t = 85$). The results of such a summation, using mortality tables for $P(t,t_o)$ and including the exponential falloff, are presented in Table 9-1. It is seen (second column) that for a one-year exposure at age 30 the lifetime risk is 1.8×10^{-4} per WLM, or about 7.2×10^{-4} for the 4 WLM exposure considered in the example above.

As far as indoor radon is concerned, exposures continuing over the full lifetime should be considered, rather than exposures for a single year. The consequences of lifetime-exposure are shown in column four of Table 9-1, obtained by summing over the effects of each year's exposure over the lifetime. This double-summation, over years during which one might suffer cancer as well as over years of exposure, gives the lifetime risk for lifetime exposure.

It is seen from Table 9-1 that the total calculated risk of radon-induced lung cancer for an annual exposure of 1 WLM per year starting at age 1 is 0.0091, or 0.91%. In slightly different words, a person who is exposed to 1 WLM per year for every year of his life, has a 0.91% chance of suffering a radon-induced lung cancer. Almost all lung cancers are fatal, so this can be taken to represent the approximate lung cancer mortality rate.

Risk can also be expressed in terms of cancers per pCi/ℓ of radon in air. Commonly, it is assumed that the equilibrium factor is $F = 0.5$ and that there are about 40 "working months" per year. With these assumptions, 1 pCi/ℓ corresponds to 0.005 WL (see Table 2-3) and to

0.2 WLM per year. Thus, to find the effect of 1 pCi/ℓ of radon, one might expect that the numerical entries in the left side of Table 9-1 would be multiplied by 0.2. Instead, in NCRP Report No. 78, 1 pCi/ℓ is taken to be equivalent to about 0.38 WLM per year, corresponding to multiplying the numerical entries for 1 WLM by about 0.38. This higher factor stems primarily from using a higher equilibrium factor, F = 0.72, and assuming exposure for the full number of hours in the year, equivalent to 52 "working months" per year. Then 1 pCi/ℓ corresponds to 0.72 x 0.01 x 52 = 0.37 WLM/year.

TABLE 9-1. *Lifetime lung cancer risk due to radon exposure, for different duration periods, as calculated in model of NCRP Report No. 78.*

Age at first exposure	Exposure at rate of 1 WLM per year (Multiply by 10^{-4})			Continuous exposure to 1 pCi/ℓ (Multiply by 10^{-4})		
	1 year	10 years	lifetime	1 year	10 years	lifetime
1	0.64	7.7	91	0.25	2.9	36
10	0.91	11	91	0.36	4.2	35
20	1.3	15	77	0.50	5.8	30
30	1.8	21	77	0.71	8.1	25
40	2.1	20	45	0.83	7.5	17
50	1.7	14	27	0.67	5.6	10
60	1.3	9.1	13	0.48	3.6	4.8
70	0.70	3.8	3.8	0.27	1.5	1.5

Reference: NCRP 1984b, Tables 10.2 and 10.3.

1. Risk calculation for left-hand set based on conversion factor for environmental conditions: 0.7 rad per WLM.
2. Risk calculation for right-hand set based on conversion factor for environmental conditions: 0.27 rad/year per pCi/ℓ of radon. This corresponds to an equilibrium factor of approximately $F = 0.72$, and a ratio of unattached RaA concentration to radon concentration of 0.07.

In summary, the calculated chances of dying from lung cancer due to radon exposure, assuming constant exposure over the entire lifetime, are:

Radon daughter exposure of 1 WLM/year	0.91%
Radon concentration of 1 pCi/ℓ (taken equal to 0.2 WLM/yr)	0.18%
Radon concentration of 1 pCi/ℓ (taken equal to 0.4 WLM/yr)	0.36%

To estimate the cancer rate in the United States due to radon, it is necessary to know the average exposure to radon daughters. It has been somewhat conventional to assume a concentration of 1 pCi/ℓ and an equilibrium factor of 0.5 corresponding, as above, to an annual exposure of about 0.2 WLM. However, for single-family houses the arithmetic mean concentration is approximately 1.5 pCi/ℓ (see Chapter 5). For purposes of estimating consequences, we take 1.5 pCi/ℓ for the mean indoor radon concentration, and take for the equilibrium factor, $F = 0.5$. This implies an exposure level of 0.75 WL and, for the equivalent of 40 "working months" per year, an average annual exposure of 0.3 WLM per year.

If this exposure is continued for a lifetime, the NCRP model implies an individual lung cancer risk of $0.3 \times 9.1 \times 10^{-3}$, or 0.27% (see Table 9-1). The annual deaths from this exposure for the U.S. population of 240 million can be very roughly estimated by multiplying the individual probability by the number of people, and dividing by an average lifespan of about 70 years, to give a time averaged impact:

$$0.0027 \times 2.4 \times 10^8/70 = 9,000 \text{ deaths per year}$$

A similar value follows from a more detailed calculation in which the total deaths are found by calculating the mortality rates as a function of age and summing over the U.S. age distribution.

It should be noted that this value, of about 9,000 lung cancer deaths per year, represents only one of a broad range of possible estimates. Many features of the calculation are open to substantial question, including the interpretation of the experience of miners, the form of the model for relating risk to radon exposure, and the choice of values for the parameters which determine the magnitude of the exposure. Some alternative results will be discussed briefly later in this chapter.

C. Absolute and relative risk

1. Basic terminology. Two broad classes of models are in common use for estimating the consequences of radiation: the absolute risk and the relative risk models. The NCRP calculation, described in Section B, was based on an absolute risk model (modified to include some dependence on age). However, as indicated in Chapter 8, relative risk models may be equally appropriate for estimating the lung cancer risk, in particular if there is a synergism between radon and smoking.

The distinction between the models can be illustrated with a specific example. Consider two populations of equal size, Populations A and B, which differ in smoking habits or in average age (or both). Assume that in the absence of excess radiation the lung cancer mortality rates are 1000 deaths per year for Population A and 2000 deaths per year for Population B. Suppose that 10 units of radiation increase the mortality rate in Population A from 1000 to 1500 deaths per year. In the absolute risk model, this means that 10 units of radiation cause 500 additional deaths and that for Population B there will be an increase from 2000 to 2500 deaths per year. In the relative risk model, 10 units of radiation cause a 50% increase in the death rate, implying for Population B an increase from 2000 to 3000 deaths per year.

The relative risk model can be used to take into consideration the dependence of cancer rate on age and, for lung cancer, on smoking. The cancer rate is higher for smokers than for non-smokers and therefore in the relative risk model the impact of radon is greater upon smokers than upon non-smokers. The impacts are equal in the absolute risk model.

2. Comparison of risks for smokers and non-smokers. If the existence of a synergism between radon and smoking is itself uncertain, little confidence can be placed in the quantitative accuracy of results obtained from any specific synergistic model. Nonetheless, consideration of specific models makes it possible to see the character of the differences between relative and absolute risk models. Broadly speaking, for the population as a whole the two models predict roughly the same total number of lung cancers, but the comparative impact of radon on smokers vs. non-smokers (and on the middle-aged vs. the elderly) is very different.

As discussed in Section B, the main NCRP calculation used an absolute risk approach and found a lifetime risk of 0.0091 for a lifetime exposure of 1 WLM per year. The NCRP study briefly reported the results of a relative risk calculation, based on a parameterization of a synergism between cigarette smoking and radon suggested by Whittemore and McMillan (1983). The calculated risk for the population as a whole was virtually the same as in the absolute risk calculation: a lifetime risk of 0.009 for a lifetime exposure to 1 WLM per year.

A detailed calculation of relative and absolute risks has been presented by Ginevan and Mills (1986). For the population as a whole, the lifetime lung cancer risk for a lifetime exposure of 1 WLM per year was again found to be about the same in the two models: 0.026 in the relative risk model and 0.021 in the absolute risk model. These results average about 2.5 times the values found in NCRP Report No. 78, but it is not clear how much significance to place on the difference. The emphasis in this study was on the interaction between cigarette smoking and radon, and some of the differences between this study and the NCRP study, such as the apparent omission of the exponential decrease factor, may have been motivated by considerations of simplicity rather than by an intended disagreement with the NCRP analysis.

In this relative risk model, the radon risk for male smokers is more than ten times the risk for non-smokers and eliminating smoking would reduce the overall impact of radon by roughly a factor of five. Very crudely, one can then think of 20% of the "radon problem" as being due to radon alone and 80% as being connected to smoking. Were this a well-established result, it could have interesting implications for radon protection standards, but it appears premature to assume synergism to be firmly established, let alone place confidence in such quantitative estimates. Thus, the extent of synergism between radon and smoking remains an important, unresolved question.

D. Lung cancer risk as calculated in other studies.

A Canadian study (Thomas 1985) has examined much the same data on miners as considered in NCRP Report No. 78, but adopted a somewhat different approach and reached higher risk estimates. The authors suggest that the difference stems in part from the possibility that in the U.S. studies of Colorado miners "exposure levels had been systematically overestimated". This would account for the lower risks per WLM found in the U.S. studies than found in other studies (see Table 8-2).

The Canadian analysis specifically addressed the consequences of lifetime exposure to 0.02 WL—the level above which remedial action should be taken, according to the policies of the Atomic Energy Control Board of Canada. It was assumed that the 0.02 WL exposure occurred 17 hours per day, seven days a week, corresponding to 36.5 "working months" per year and a total exposure of 0.73 WLM per year. At this level of exposure, averaging the results of several absolute and relative risk models, the study concluded that there will be "20 excess lung cancers per 1000 persons".

This result converts to a probability of 0.027 (2.7%) of suffering lung cancer for a lifetime exposure of 1 WLM per year—an impact three times greater than that found in NCRP Report No. 78. Assuming linearity, and again taking an annual exposure of 0.3 WLM per year, this corresponds to roughly 28,000 deaths per year in the U.S.

The difficulties faced in obtaining estimates of total radon impact are illustrated by the numbers given in the EPA's publication (1986): *A Citizen's Guide to Radon.* In its most conspicuous and widely cited estimate, the annual lung cancer mortality due to radon is put at 5,000 to 20,000. However, in a table in the same publication, it is indicated that a lifetime exposure to a radon concentration of 1 pCi/ℓ implies a lifetime risk of between 0.003 and 0.013. This corresponds to roughly 10,000 to 45,000 deaths per year in the U.S. Taking into account the fact that the average radon concentration is probably higher than 1.0 pCi/ℓ this range could be put still higher.

It should be noted that the EPA estimates are based on relative risk models. Thus, even were these numbers to be accepted, the estimated lung cancers cannot be attributed solely to radon but involve smoking as well.

E. Summary

The existence of the high estimates cited in Section D makes understandable the conclusion for indoor radon reached by the Federal Interagency Committee on Indoor Air Quality, co-chaired by representatives of the Environmental Protection Agency and the Department of Energy, that "the estimated upper limit of health effects is 30,000 deaths per year" (CIAQ 1985). However, there are several reasons to question so high a limit, even if it cannot be firmly excluded.

For one, this figure follows from assuming a high rate of cancer incidence per WLM and exacerbates the problem of understanding the results for United States uranium miners, especially at low dose rates (see Chapter 8). The seriousness of the contradiction depends on the confidence to be placed in the analyses of these data. Thomas *et al.* (1985) have suggested as limitations of the studies of U.S. uranium miners "the poor quality of the exposure data and the small number of subjects exposed to low levels." It is not clear whether it will be possible with further study to develop sufficient confidence in the information on miners, especially at relatively low dose rates, to rule out the high estimates for radon impact on the general population.

Other arguments against a limit as high as 30,000 radon-induced cancer deaths per year hinge upon whether one accepts an absolute risk model or a relative risk model. The analysis is much simpler in an absolute risk model, although it is generally deemed unlikely that a strict absolute risk model is correct. Assuming an absolute risk model, in which the risks are purely additive, a limit can be set on radon-induced lung cancer deaths by comparing total deaths and those attributed to smoking. The American Cancer Society estimates that 83% of the 136,000 annual lung cancer deaths are due to smoking (ACS 1987). If these numbers are taken literally, with no allowance for inherent uncertainties or for synergistic effects, this would leave only 23,000 deaths from other causes and the estimated 30,000 deaths from radon would exceed the total non-smoking component.

It is in principle possible to address the issue directly on the basis of epidemiological studies of lung cancer incidence in regions with high indoor radon levels, although again existing studies are not conclusive. As discussed in Chapter 8, the predicted increase in lung cancer deaths is not seen in studies made to date. The predictions correspond to intermediate estimates of radon impact, not to high estimates such as represented by a figure of 30,000 deaths per year. Thus, current epidemiological studies for the general population suggest possible problems in accepting a central estimate of, say, 10,000 deaths per year and still greater problems for any of the higher estimates.

The analysis is more complicated if one adopts a relative risk model or any intermediate model in which there is a substantial synergism between radon and smoking. A simple division of the lung cancer deaths into a smoking component and a radon component is then impossible. Epidemiological studies of the general population are still in principle possible, but they are more complex. For a population in which there is little smoking, as was the case for the Cumberland County study (Cohen 1987) and possibly the Guangdong study (Hofmann 1986), the calculated lung cancer increase at elevated radon concentrations could be small if the synergism is strong. Thus one would learn little about the impact of elevated radon levels on an "average" population, which includes a substantial number of smokers.

However, the terminology also becomes more complicated if there is a strong synergism. It would not be appropriate to say that "radon causes 30,000 lung cancer deaths" if the number comes from a relative risk model in which smoking is also strongly implicated. The "radon deaths" would be far fewer were people to adopt the simple remedial measure of ceasing to smoke, and the ascription of cause should include both factors.

The discussion above has concentrated on the upper end of the risk estimates. There are questions about the lower end as well. Again with the caveat that the studies are not conclusive, the absence of compelling direct evidence of increased lung cancer rates in regions of high indoor radon concentration raises doubts about the validity of the linearity hypothesis, which is basic to all the standard estimates including the EPA's lower bound of 5,000 deaths per year. Cohen (1985) has suggested that the "straightforward explanation... is that the linear dose relationship for radiation carcinogenesis *grossly* overestimates the effects of low level radiation, even for alpha particles..." If Cohen is correct, then the number of cancers produced by indoor radon could be well below the estimate of 5,000, cited here as a low estimate.

In conclusion, it is not possible to specify with any certainty at this time the number of lung cancer deaths which are being caused in the U.S. by indoor radon. For regulatory and planning purposes, it is reasonable to speak in terms of 10,000 deaths per year as a nominal measure of radon's impact and such a number is commonly cited. The main advantages of this nominal round number is that figures of approximately this magnitude follow from current estimates of average radon concentrations together with conservative models for cancer induction by radon. However, both the magnitude of the effects and even the meaning of the term "due to radon" contain large ambiguities.

References: Chapter 9

American Cancer Society, 1987, *Cancer Facts and Figures—1987* (New York: ACS).

Interagency Committee on Indoor Air Quality, 1985, "Report of the CIAQ Radon Workgroup," R. J. Guimond and W. Lowder, co-chairs, Radon Working Group, May, 1985 (Washington, D.C.: CIAQ).

Cohen B. L., 1985, "Lung Cancer in High Radon Areas and Its Implications for the Cancer Risk of Low Level Radiation," (unpublished manuscript).

______, 1987, "Tests of the linear-no threshold dose-response relationship for high-LET radiation," *Health Phys.* **52**, 629.

EPA (Environmental protection Agency), 1986, published with The Department of Health and Human Services, *A Citizen's Guide*

To Radon, What It Is and What To Do About It, OPA-86-004 (August).

Ginevan M. E. and Mills W. A., 1986, "Assessing the risks of Rn exposure: the influence of cigarette smoking," *Health Phys.* **51**, 163.

Harley N. H. and Pasternack B. S., 1981, "A model for predicting lung cancer risk, induced by environmental levels of radon daughters," *Health Phys.* **40**, 307.

Hofmann W., Katz R. and Zhang Chunxiang, 1986, "Lung cancer risk at low doses of α particles," *Health Phys.* **51**, 457.

NCRP (National Council on Radiation Protection and Measurements), 1984, "Evaluation of Occupational and Environmental Exposures to Radon and Radon Daughters in the United States," *NCRP Report No. 78* (Bethesda, MD: NCRP).

Thomas D. C., McNeill K. G. and Dougherty C., 1985, "Estimates of lifetime lung cancer risks resulting from Rn progeny exposure," *Health Phys.* **49**, 825.

Whittemore A.S. and McMillan A., 1983, "Lung cancer mortality among U.S. uranium miners: a reappraisal, " *National Cancer Inst.* **71**, *489.*

Chapter 10

COMPARISONS OF INDOOR RADON TO OTHER RADIATION HAZARDS

David Bodansky, Kenneth L. Jackson, and Joseph P. Geraci

A. Comparison of radon to other causes of cancer

The significance of radon as a cause of cancer can be put into perspective by a comparison with other causes of cancer. Lung cancer is fatal in about 90% of the cases and therefore in assessing radon exposures it is pertinent to make a comparison in terms of mortality figures. As discussed in Chapter 9, there is a broad range of estimates for the U.S. lung cancer mortality from indoor radon, extending from well below 5,000 to above 20,000 per year. For simplicity in making comparisons, we will use here an intermediate estimate of 10,000 per year, although one needs to bear in mind that this number is uncertain.

Given approximately 480,000 fatal cancers in the United States per year and 136,000 fatal lung cancers (see Chapter 8), this estimate corresponds to indoor radon being responsible for 2% of all cancer mortality and 7% of lung cancer mortality. It is estimated that 83% of lung cancers are due to smoking (ACS 1987), corresponding to about 113,000 deaths per year. Overall, therefore, smoking is far more important than indoor radon as a cause of lung cancer. Despite our concentration on it, radon is a relatively small contributor in the perspective of cancer as a whole or even of lung cancer alone. As will be seen in the succeeding sections, it is a more dominant contributor in considering average exposures to ionizing radiation.

B. Calculation of effective radiation doses from radon

1. Dose comparisons vs. mortality comparisons. The analysis of the impact of radon has been couched in Chapters 8 and 9 in terms of lung cancer mortality but, as discussed above, the mortality rates are highly uncertain. Another perspective may be obtained by comparing the magnitude of the radiation doses caused by radon to the magnitude of the doses from other sources of ionizing radiation.

By making a comparison based on radiation doses, one avoids the need for an explicit assumption about the effects of radiation at low doses. More specifically, one avoids the extrapolation from the lung cancer mortality rate for miners (at high dose levels) to expected deaths among the general population (usually at much smaller dose levels). A linear dose-response curve is usually assumed in making this extrapolation, but if radiation has a much reduced effect (or no effect) per unit dose at low dose levels, then it is wrong to apply a linear extrapolation based on the experience at the high dose levels where the induction of cancer has been observed (see Chapter 8).

There are specific procedural difficulties in making a comparison in terms of doses because radon affects only a part of the lung while other radiations can affect the entire body and because alpha particles are more damaging, for a given energy deposition, than are beta or gamma rays. These complications are taken into account through the introduction of modifying factors, the weighting factor and the quality factor, discussed below. The resulting radon dose, so adjusted, is termed the "effective dose equivalent", and can be directly compared to doses received from other sources of radiation as a means of estimating overall hazard.

In the remainder of Section B, the determination of the effective dose equivalent from radon will be considered. In Section C, this dose will be compared to radiation doses encountered from other sources, and in Section D a comparison will be made between regulatory standards for radon and for other sources of radiation.

2. Regional weighting factors. In considering radiation exposures, it is necessary to distinguish between cases where only a limited part of the body is involved, as for radon inhalation, and cases where the entire body is exposed at more or less the same level. In the latter case, the dose is described as a "whole-body" dose. For the exposures received from cosmic rays, which pass through all parts of the body equally, it is clearly correct to speak in terms of a whole-body dose. In other cases, such as the exposures received from gamma-ray emitters in the ground, treating the received dose as a whole-body dose gives a reasonably close approximation to the actual situation.

When the exposure is primarily to a single organ, radiation effects can still be put on a common footing by considering an "effective" whole-body dose. The relation between the dose to the individual organ and the effective whole-body dose is expressed through a so-called weighting factor, sometimes also called the regional dose factor. More specifically, the weighting factor for tissue T, W_T, is defined as "the proportion of the stochastic risk resulting from tissue (T) to the total risk, when the whole body is irradiated uniformly" (ICRP 1977a, p. 21). Here, the term "risk" refers to the risk of any radiation-induced fatal cancer plus any genetic damage expressed in the next two generations. The meaning and use of the weighting factor is best explained through an example. Suppose the lung receives a dose of 1000 millirad and the weighting factor for the lung is $W_L = 0.12$. Then the effective whole-body dose is $0.12 \times 1000 = 120$ millirad.

The choice of weighting factor for converting from the lung dose to the effective whole-body dose has been discussed in the ICRP Publication 32 of the International Commission on Radiological Protection (1981). Two alternatives are presented, the so-called "mean lung dose-concept" and the "regional lung dose-concept." In the mean lung dose-concept, the lung is considered as a single organ and the average dose is used. For this model, a weighting factor W = 0.12 is recommended. In the regional lung dose-concept, doses to the basal cell layer of the tracheo-bronchial region and the pulmonary regions are considered separately, with equal weighting factors recommended in ICRP Publication 32: $W_{T-B} = W_P = 0.06$. The calculated effective dose depends sensitively on the choice of the weighting factor for the tracheo-bronchial region, because radon daughter deposition is greater there than in the pulmonary region. Thus, were one to adopt a value for W_{T-B} greater than 0.06, the effective whole-body dose would be increased.

3. The quality factor. As discussed in Chapter 2, the same physical dose (in rad or gray) can lead to differences in the severity of the biological effects, depending upon the density of ionization of the particles involved. In considering the rate of cancer induction by different types of ionizing radiation, the useful dose is the "dose equivalent" (in rem or sievert), found by multiplying the physical dose by a scale factor, called the quality factor (Q). At low doses, the dose equivalent is commonly expressed in terms of millirem (mrem) or milli-sievert (mSv), where 1 rem = 1000 millirem = 10 mSv.

In recent years, it has become common practice to adopt as a quality factor for alpha particles, Q = 20. This value was adopted in 1977 by the ICRP in its Publication 26 (ICRP 1977a) and is used in most current analyses. For example, this was the assumed value in the analyses in a 1983 study by an expert panel of the European OECD (NEA 1983), it was employed in the NRCP Report No. 77 (NCRP 1984a, p. 80) for the conversion of radon doses, and it was "the currently recommended value" in a recent review of radon risks (Cross 1985). However, some uncertainty remains, as reflected in Report No. 78 of the NCRP (1984b, p. 40), where values from 3 to 20 are mentioned, with the admonition that the issue "needs addressing". The Report of the Committee on the Biological Effects of Ionizing Radiation (BEIR III) estimates, with "substantial" uncertainties, a relative biological effectiveness for alpha particles of between 8 and 15 (NAS 1980, p. 327). We will here follow the conventional practice of using Q = 20 but significantly lower, and possibly higher, values cannot be excluded.

4. Calculation of effective dose equivalent per WLM. The "effective dose equivalent," which is the appropriate dose to use in comparisons of the overall effects of doses from different sources of exposure, is found by multiplying the physical dose to the individual organ by the two correction factors cited above. More specifically, for radon, the effective dose equivalent per WLM is given by the relations:

$$H_E = Q \times D \times W \qquad (1)$$

or

$$H_E = Q \times [D_{T-B}W_{T-B} + D_P W_P] \qquad (2)$$

where:

H_E= effective dose equivalent (e.g., in mrem per WLM)
Q= quality factory for alpha particles
D= dose conversion factor for specific organ (e.g., mrad per WLM)
W= weighting factor for specific organ

In expression (1), no distinction is made between different parts of the lung. In expression (2), the subscripts T-B and P refer to the tracheo-bronchial and pulmonary regions, following a distinction made in some analyses.

From expression (2), using the values for the weighting factor and quality factor discussed above, it follows that

$$H_E = 20 \times [0.06 \times D_{T-B} + 0.06 \times D_P] = 1.2[D_{T-B} + D_P] .$$

The dose conversion factor, D_{T-B}, has been calculated in several models, as discussed in Chapter 7 (NEA 1983, NCRP 1984b). These calculations differ in the parameters used to describe conditions for the typical indoor environment. For purposes of making a "best" estimate, the result from NCRP Report No. 78 is used here: D_{T-B} = 710 mrad/WLM. The NCRP Report does not give a value for D_P. Using the data from Table 7-1 for the NEA model, it can be can inferred that D_P is roughly one-sixth of D_{T-B} : D_P = 120 mrad/WLM. With these dose conversion factors, one obtains the approximate result: H_E = 1000 mrem per WLM (10 mSv per WLM). A number of other estimates are presented in Table 10-1.

Noting both the extremes in the estimates in Table 10-1 and also the inherent uncertainties in the calculations which can be made at this time, the effective dose equivalent for inhalation of radon daughters is best specified in terms of a range of possible values:

$$H_E = 500 - 2000 \text{ mrem per WLM } (5 - 20 \text{ mSv/WLM})$$

As a rough central estimate, 1000 mrem per WLM (10 mSv per WLM) is adopted here, but this should be viewed as a nominal number cited for specificity in discussion. It is uncertain by at least a factor of two.

5. Consistency check: effects of 1 WLM/yr and 1000 mrem/yr. A rough consistency check can be made by comparing the estimated effects of a lifetime radon exposure of 1 WLM per year, as extrapolated from the experience of miners, and the effects of a lifetime whole-body dose of 1000 mrem per year.

As discussed in Chapter 9 and indicated in Table 9-1, the NCRP interpretation of the data for miners suggests that a radon exposure of 1 WLM per year corresponds to an 0.91% lifetime risk of fatal cancer. The effects of a dose of 1000 mrem per year, are estimated in the BEIR III Report (NAS 1980, p. 146) for different dose-response models. To compare to the results of the NCRP model, it is appropriate to use the figures for a linear dose-response, absolute risk model. For this case, BEIR III estimates a 1.12% mortality risk for a lifetime dose

TABLE 10-1. *Dose conversion factors cited or implied in literature: effective equivalent whole-body dose per WLM exposure to radon-222 daughters.*

Source	Reference	mrem/WLM	Note
NCRP (Report No. 77)	NCRP 1984a, p. 80	1000	(a)
NCRP (Report No. 77)	NCRP 1984a, p. 94	2000	(b)
NEA-OECD	NEA 1983, p. 65	550	(c)
Cohen	Cohen 1985	540-1800	(d)
James	James 1987	1500	
This report		1000	(e)

Notes:

a. From equivalence attributed to the ICRP: 0.5 WLM/yr = 500 mrem/yr.

b. Derived from NCRP estimate that the number of cancer deaths caused by an exposure of 2 WLM/yr is eight time greater than the number caused by 500 mrem/yr of external gamma-ray radiation.

c. The tabulated number is the proposed reference level. A range from 490 mrem/WLM to 1090 mrem/WLM is cited, for different assumptions as to ventilation (i.e. unattached fraction) and lung model.

d. Calculation based on input data from NCRP, UNSCEAR, and ICRP. Using the ICRP data alone, Cohen finds a conversion factor of 540 mrem/WLM, in agreement with the NEA-OECD value. Relying more on the NCRP data, he finds a factor of 1800 mrem/WLM.

e. See Section B-4; the range is there given as 500-2000 mrem/WLM.

of 1000 mrem per year. Thus, the estimated effects are the same to within about 20%.

Alternatively, these figures could be used to calculate the effective dose equivalent per WLM. The result, 810 mrem per WLM [(0.91/1.12) x 1000 = 812], is well within the previously estimated band of 500 to 2000 mrem per WLM and is consistent, within the uncertainties of the estimates, with the adopted conversion factor of 1000 mrem per WLM.

6. *Effective dose equivalent from indoor radon.* A common estimate of the mean radon concentration for single-family houses is 1.5 pCi/ℓ (see Chapter 5). Noting that about 70% of the housing units are single-family units (DOE 1984) and taking the radon concentration to be about 0.5 pCi/ℓ in multi-family dwellings (Nero 1986), this would correspond to an average radon concentration for the entire housing stock of about 1.2 pCi/ℓ. However, it is probably unreasonable to make this small correction. In extensive recent measurements of radon concentrations (Alter 1987) the average concentration was found to be far above 1.5 pCi/ℓ. While these data sets may be unrepresentative, in that the measurements were made on the initiative of the homeowners and thus may be biased to favor localities with high radon concentrations, they raise serious questions as to the true

average radon concentration. It appears reasonable to adopt tentatively a concentration of 1.5 pCi/ℓ for the housing stock as a whole, in a crude compromise between adjustments which might be made in either direction. This concentration corresponds to a mean exposure level of about 0.3 WLM per year (assuming an equilibrium factor of 0.5 and 40 "working months" per year).

Allowing for the large uncertainties in these numbers, a range of plausible estimates for the key parameters is:

conversion factor (mrem per WLM)	500 to 2000
average exposure (WLM per year)	0.2 to 0.4

This implies that the average effective dose equivalent from indoor radon lies between 100 mrem per year and 800 mrem per year. A single central estimate can be obtained by taking the dose conversion factor to be 1000 mrem per WLM and the average exposure in the U.S. to be 0.30 WLM per year. This gives a central estimate for the effective dose equivalent of: 300 mrem per year. Thus we conclude the average annual effective dose equivalent in the U.S. from indoor radon to be:

possible range: 100 to 800 mrem	(1 to 8 mSv)
central estimate: 300 mrem	(3 mSv)

The central estimate is used below as a nominal average value on which to base further discussion and comparisons.

C. Comparison of doses from radon to doses from other sources

1. Routine exposures. We here compare the dose from indoor radon to radiation doses from other sources. A summary of doses from a variety of sources is presented in Table 10-2, based primarily on data in the BEIR III Report (NAS 1980). As seen in Table 10-2, the total dose from all sources other than medical irradiations and radon is about 100 mrem per year. The doses from medical irradiations vary greatly with the individual, but the population average was indicated in BEIR III to be about 100 mrem per year. Thus, if the average effective dose equivalent from indoor radon is 300 mrem per year, then radon produces an average population dose which exceeds the dose produced by all other sources combined.

2. Comparisons to nuclear accidents. Another perspective on the relative importance of radon exposures can be gained by comparing them not to routine situations but to extreme cases — in particular to exposures resulting from nuclear reactor accidents.

TABLE 10-2. *Annual whole body equivalent doses in the United States from major radiation sources.*

Source	Ref.	Exposed Group	Dose (mrem) Group	Pro-rated	
Natural radiation					
cosmic radiation	NAS80	whole population	28	28	
terrestial radiation	NAS80	whole population	26	26	
internal sources	Ha83	whole population	40	40	(a)
Weapons tests	NAS80	whole population	5	5	
Airplane travel	NAS80	passengers	3	1	(b)
Nuclear reactors	NAS80	people within 10 mi	$\ll$ 10	$\ll$ 1	(c)
	Ja83	people within 50 mi	0.01	0.01	(d)
Consumer products	NAS80		—	4	
Medical exposures	NAS80		—	100	(e)
TOTAL				200	(f)
Indoor radon		whole population		300	(g)

Notes:

a. Approximate average of doses for different organs (as high as 105 mrem for bone surfaces). The values here are higher than those in BEIR III because a quality factor of 20 was adopted for alpha particles.

b. Average dose for passengers group based on 15 hours per year at altitude of 9.5 km (0.2 mrem/hour).

c. The indicated upper limit is based on regulatory requirements; the actual exposures are much lower.

d. The average annual dose for people within 80 kilometers of nuclear reactors was 0.007 mrem in 1979, according to Nuclear Regulatory Commission estimates (Jacobs 1983).

e. Rounded-off value of the sum (92 mrem) of each of the contributions to the pro-rated population dose, as listed in NAS 1980, Table III-23 (based primarily on 1970 studies). Individual doses vary widely.

f. Rounded-off value of the sum of the entrees above.

g. Estimated average dose for the entire population; alternative estimates for the average range from 100 to 800 mrem per year (see Section B-6). Individual doses vary widely.

Such a comparison can be made for the two worst-case events to date in the United States, which by chance have both been in Pennsylvania. The maximum individual exposure to a person in the vicinity of the Three Mile Island nuclear reactor at the time of the 1979 accident (excluding plant workers) was 70 mrem and the average exposure for the two million people living within 50 miles was about 1 mrem (Kemeny 1979, p. 34). For people exposed to radon in the Reading Prong region of Pennsylvania (see Chapter 5) the maximum level was over 10 WL, corresponding to an effective dose equivalent of over 400,000 mrem if continued for a year; the average exposure for a sample of 5011 homes was 0.078 WL, corresponding to an effective dose equivalent of about 3000 mrem per year (Gerusky 1986).

Thus, whether one compares the extreme exposures or average exposures, the Reading Prong radon exposures were over a thousand times greater than those from Three Mile Island. Of course, the radiation impact of Three Mile Island is known to have been very small, reducing the significance of a comparison to it.

Less is known at this time about the far more serious 1986 accident at Chernobyl in Russia. A Soviet report issued in December 1986, eight months after the accident, indicated that the accident had caused 31 deaths (Nuclear News 1987). In terms of the immediate injuries and fatalities, the disruption of living patterns, and the need for drastic protective measures the consequences are much more severe than any from indoor radon exposures. In contrast to Chernobyl, there have been no acute deaths from radon exposure. However, in terms of eventual cancer mortality the comparison between Chernobyl and radon comes out differently.

Preliminary estimates of the effects of Chernobyl can be made on the basis of the information released by the Russian authorities at a meeting of the International Atomic Energy Agency in Vienna in August, 1986. The total dose from external radiation received by the 74.5 million people in the Western USSR due to Chernobyl was estimated to be 8.6 million person-rem for 1986 and 29 million person-rem for the full 50-year period following the accident (USSR 1986, Annex 7, p. 62). This corresponds to average individual doses of about 110 mrem in 1986 and 8 mrem per year averaged over the 50-year period. It is now commonly estimated that internal radiation from ingestion of contaminated foods will cause a roughly equal radiation exposure. Thus, including the effects of both external and internal radiation, the average individual long-term radiation dose for this population is approximately 16 mrem, which is less than one-tenth of the estimated average effective dose equivalent from indoor radon in the U.S.

Estimates of cancer fatality rates from Chernobyl have been commonly made, in the interests of conservatism, using the linearity hypothesis. As discussed in Chapter 8, this hypothesis is judged in BEIR III (NAS 1980) probably to give an overestimate of the impact of beta-particle and gamma-ray exposures, which are responsible for most of the Chernobyl dose. Assuming a linear dose-response curve and using conventional risk estimate figures (NAS 1980, p. 145), the external exposure is calculated to lead to about 5000 fatal cancers. Including internal radiation, the estimated total cancer toll from Chernobyl for the Western USSR over the next 50 years is therefore about 10,000 deaths. During the same period, a population of 74.5 million in the United States would be expected to suffer approximately 150,000 fatal cancers from indoor radon, again assuming linearity. It may also be noted that if Russian cancer mortality rates are similar to current U.S. rates, the Chernobyl toll of 10,000 possible cancers will be submerged in over 7,000,000 cancer deaths in the Western USSR from other causes. With so small a fractional impact, it will be very difficult to observe the long-term effects of Chernobyl except in special areas or to judge whether or not these calculations have provided a correct estimate of these effects.

3. Difficulties with dose comparisons. The comparisons presented above for radon exposures rely on estimates of the physical lung dose, the weighting and quality factors, and on the average radon daughter concentrations. Aside from the fact that each of these is an empirical term, obtained from not fully adequate observational data, there are other difficulties of a more general sort.

One difficulty arises in the use of the empirical weighting and quality factors. The weighting factors are based on observations of human cancer incidence and on estimates of genetic risk at high dose levels. Quality factors are based on data from both human epidemiological studies and animal experiments, looking at a variety of effects, again at high dose levels. There is no assurance that the values found for these parameters remain the same at the lower levels characteristic of indoor radon exposures.

A further caveat in interpreting these comparisons relates to the possible different effects at different ages. In the NCRP model (NCRP 1984b), it is assumed that no lung cancers occur before age 40 and that repair processes or removal of malignant cells reduce the effects at age 40 (and beyond) of exposures received at an early age (see Chapter 9). However, for a given radon exposure the dose is higher for children than for adults due to differences in lung size and breathing rates (i.e. the conversion factor from WLM to rads or millirem is higher) (NCRP 1984b, p. 48). These effects act to cancel, and in the NCRP model make radon exposures in childhood less damaging than those received past age 40. Nevertheless, they serve as reminders of age-dependent impacts and it may be noted that, in contradiction to implications of the NCRP analysis, the Environmental Protection Agency (1986) raises the possibility that "children could be more at risk than adults from exposure to radon."

It also must be noted that radon is important only for its somatic (i.e. cancer producing) effects while whole-body radiation exposures may cause both somatic and genetic damage. As pointed out, for example, in the BEIR III Report "it is extremely difficult to compare the societal impact of a cancer with that of a serious genetic disorder" (NAS 1980, p. 6). Nonetheless, it is generally believed that the adverse genetic consequences of whole body exposure, including the effects on future generations, are somewhat less than the adverse somatic effects. Thus, the International Commission on Radiological Protection estimates (ICRP 1977a, par. (43)) that "the hereditary detriment is likely to be less than the detriment due to somatic injury." An expression of this prevailing belief is implicitly contained in radiation protection guides, which do not establish more stringent limits on irradiation of the gonads than on irradiation of the remainder of the body. In any event, in establishing the weighting factors for the lung, the ICRP has already attempted to allow for genetic effects. Were these ignored, the lung weighting factor and effective whole body dose from radon exposure would be greater.

Overall, these difficulties do not invalidate the comparisons made between radon exposures and exposures from other radiation sources, but they suggest that such comparisons are to be taken as approximate guides rather than as predictors of precisely determined relative impacts.

D. Radiation protection standards

1. General considerations. There has been concern over the possible adverse effects of ionizing radiation since shortly after the original discoveries of x-rays and of radioactivity near the turn of the century, but in the early years routine practices were uncontrolled and often highly dangerous. A major step in the codification of the international scientific community's response to radiation hazards came in 1928 with the formation of the ICRP. This organization remains the international leader in the promulgation of recommendations as to acceptable levels of radiation exposure.

The current guiding viewpoint toward radiation exposures is expressed in a series of general recommendations for dose limitation enunciated by the ICRP in its Publication 26 (ICRP 1977a). In addition to quite specific recommendations, this Publication states the following key principles for the setting of standards (ICRP 1977a, par. 12):

a) no practice shall be adopted unless its introduction produces a positive net benefit;

b) all exposures shall be kept as low as reasonably achievable, economic and social factors being taken into account; and

c) the dose equivalent to individuals shall not exceed the limits recommended for the appropriate circumstances by the Commission.

The criterion "as low as reasonably achievable" is commonly referred to by the acronym ALARA. As discussed below, the allowance for "economic and social factors" is perforce much greater for indoor radon than for other sources of (non-medical) radiation.

Individual countries have established their own organizations, both for the development of recommendations and the establishment of regulations. In the United States, recommendations are made by the National Council on Radiation Protection and Measurements (NCRP) and regulations with the force of law are established by the Environmental Protection Agency (EPA) and the Nuclear Regulatory Commission (NRC). Overall, these recommendations and regulations are similar to the ICRP recommendations, both in spirit and in specific requirements.

2. Standards for sources other than radon. The ICRP recommends the following limits for annual radiation exposures, excluding normal natural radiation levels and medical exposures:

a) 5000 millirem (50 mSv) as an occupational limit for individuals who receive radiation exposures in their work. [ICRP 1977a, Par 104]

b) 500 millirem (5 mSv) for the individual members of the general public who receive the highest exposures. [ICRP 1977a, Par 119]

c) 100 millirem (1 mSv) for the average exposure of the population as a whole. [ICRP 1985]

This last recommendation, of a 100 millirem per year limit on the average exposure of the general population, was put forth recently as an expression of the "principal limit" for the general population. It is expected that this limit will be satisfied if the individual limit of 500 millirem per year is satisfied.

Overall standards in the United States conform reasonably closely to these recommendations. In addition, more conservative limits have been established by the EPA and NRC for facilities operated by the Federal government or licensed by the NRC for the generation of electricity from nuclear power. These are embodied in the Code of Federal Regulations (especially in 10CFR50, Appendix I (CFR 1986a); 40CFR61 (EPA 1985a); 40CFR190 (CFR 1986b); and 40CFR191 (EPA 1985b)). The basic requirement for these facilities is that in standard operation no facility emit radioactive material, other than radon, in amounts which will give any member of the general public a dose of more than 25 millirem per year. For some facilities (but not for commercial nuclear reactors or nuclear waste disposal sites) alternative standards allow the 25 millirem requirement to be relaxed, but only if no member of the general public will receive a dose in excess of 100 millirem per year from all sources other than natural background and medical procedures.

The EPA requirements for nuclear waste repositories not only set the 25 millirem per year limit, but require that they be satisfied for 1000 years after the original disposal of the waste. In discussing the repository regulatory criteria, including specific limits on the amounts of radioactive material which can be released, it is stated (EPA 1985b:38069):

> Disposal in compliance with the containment requirements is projected to cause no more than 1,000 premature cancer deaths over the entire 10,000-year period from disposal of all existing high-level wastes and most of the wastes to be produced by currently operating reactors— an average of 0.1 fatality per year.

3. Standards for radon. Standards for occupational exposure to radon, as in uranium mining, are close to those established for other occupational radiation hazards. The ICRP recommends an exposure limit for workers of 4.8 WLM per year (ICRP 1981). Using the ICRP conversion factor for miners of 0.010 Sv per WLM, this translates to an effective dose equivalent of 4800 mrem (48 mSV) per year, which is indistinguishable from the general occupational limit of 5000 mrem per year. Setting the limit at 5000 mrem per year results in an average dose to workers of about 500 mrem per year, which produces health risks that are somewhat less than typical hazards encountered in non-radiation "safe occupations" (ICRP 1977b). The recommended limit for miners in the United States is 4 WLM per year (NCRP 1984b, p. 164), again essentially at the same level as other occupational exposures.

The setting of indoor radon standards for the general population is complicated by the need to remain within the realm of the practical. With the average (median) residence having a radon level corresponding to an effective dose equivalent of about 200 millirem per year, and many residences at much higher levels, it would be impractical to impose limits similar to those established for other sources of radiation. As pointed out in the NCRP Report No. 77 (NCRP 1984a, p. 87):

> Setting remedial action levels based on the distribution of natural background involves considerations of both risk and practicality. A choice must be made at a level where the remedial action required is reasonable given the costs involved and other factors. It is desirable to minimize the impact by placing the remedial action level at a point where as few buildings as possible require remedial action.

These considerations led the NCRP group to recommend a limit of 2 WLM per year. The EPA, in its recently developed "A Citizen's Guide to Radon" makes a graded set of recommendations (EPA 1986a). The target is a reduction to 0.02 WL; for levels up to 0.1 WL it is recommended that action be taken "within a few years"; and for exposures between 0.1 WL and 1 WL, action is recommended "within several months". Assuming 40 "working months" per year, the 0.02 WL target translates into a goal of 0.8 WLM per year, while the "within a few months" threshold of 0.1 WL translates into an implied need for substantial concern at 4 WLM per year, the same as the occupational limit.

The recommendations of the NCRP and the EPA are summarized in Table 10-3. However, these are not statutory limits and there is no regulatory force behind the advisory recommendations. Individuals and utilities interested in encouraging conservation are not generally barred from reducing air exchange rates even in houses which exceed these levels, much less be required to take remedial action at any level.

4. Comparisons of indoor radon and other standards. As indicated above, the target for remedial action for indoor radon is at a level of 1 or 2 WLM per year. For a conversion factor of 1000 millirem per WLM (see Section B), this corresponds to levels of 1000 to 2000 millirem per year. Table 10-4 compares these targets with the U.S.

TABLE 10-3. *Levels at which corrective action is recommended to limit exposures from indoor radon.*

Source	Description of recommendation	Limit Specified	WLM/yr Equiv
EPA 1986a	desired target	0.02 WL	0.8
	action within several months	0.1 WL	4
NCRP 1984b	remedial action level	2 WLM	2

standards for other sources of exposure for the general population. It is seen that radiation doses at the recommended action levels for indoor radon are in excess of those permitted from other sources.

As indicated above, the differences stem in large part from judgments as to what is reasonably achievable, made in a climate where the regulatory bodies are encouraged to be cautiously conservative, wherever possible. However, it is important to bear in mind that there is no clear evidence that radiation doses at the level of several hundred millirem per year cause an increase in cancer incidence in humans or animals. Thus, it may prove in retrospect that some or all of these standards have been unnecessarily conservative. In particular, the stringent standards set for nuclear reactors are not intended to be put forth here as an example of a goal towards which one should strive for indoor radon.

TABLE 10-4. *U.S. standards for limitation of the exposure of the general population to radiation (in millirem per year).*

	Standard
General limit on individual exposures[1]	500
Limit on average population exposure[1]	170
Individual exposure limit for waste disposal and reactors[2]	25
Total individual exposure limit for DOE/NRC facilities[2]	100
Target for action for indoor radon exposure (approx)[2]	800 - 4000

1. Reference: NCRP 1971.
2. See text and Table 10-3.

Looking ahead, it may not be possible to sustain indefinitely a situation in which it is deemed "acceptable" to have much higher radiation exposures from radon in the home than are permitted in the immediate vicinity of nuclear power plants or nuclear waste disposal sites. At present, however, even were it granted that consistency is desirable, there is not a sufficient consensus to permit society to decide whether to resolve the inconsistency by doing more to reduce indoor radon levels or by being less fearful of radiation from other sources.

Until there is a firmer knowledge of the effects of low levels of radiation, there will not be agreement within the scientific community as to the proper course, let alone concurrence on the part of the public as a whole. It is to be hoped that recognition of the inconsistency will help spur the development of the understanding necessary for framing a responsible and consistent societal response to radiation. This would permit more rational allocation of resources for the control of radiation and lead to more sensible decisions on the costs and benefits arising from measures such as weatherization of homes, operation of nuclear reactors, or using radiation in medical treatments. Of course, this would only be one step towards achieving a broader goal of rationalizing our response to all environmental hazards.

References: Chapter 10

ACS (American Cancer Society), 1987, *Cancer Facts and Figures—1987* (New York: ACS).

Alter H. .W. and Oswald R. A., 1987, "Nationwide distribution of indoor radon measurements: a preliminary data base," *J. Air Pollution Contr. Assn.* **37**, No. 3, 227.

Code of Federal Regulations 10, Parts 0 to 199, Revised as of January 1, 1986, Part 50, App. I (Washington, DC: U.S. Govt. Printing Office).

_____, *Parts 190-399*, Revised as of July 1, 1986, Part 190 (Washington, DC: U.S. Govt. Printing Office).

Cohen B. L., 1985, unpublished preprint (private communication).

Cross F. T., Harley N. H. and Hofmann W., 1985, "Health Effects and Risks from ^{222}Rn in Drinking Water," *Health Phys.* **48**, 649.

Cross F. T., 1986 (private communication).

Department of Energy, 1984, *Residential Energy Consumption Survey: Housing Characteristics 1982*, Energy Information Administration, DOE/EIA-0314(82), p. 15

EPA (Environmental Protection Agency), 1985a, "National Emission Standards for Hazardous Air Pollutants; Standards for Radionuclides; Final Rules," 40 CFR Part 61, *Federal Register* **46**, 5190 (February 6, 1985).

_____, 1985b, "Environmental Standards for the Management and Disposal of Spent Nuclear Fuel, High-Level and Transuranic Radioactive Wastes; Final Rule," 40 CFR Part 191, *Federal Register* **50**, 38066 (September 19, 1985).

_____, 1986, published with the Department of Health and Human Services, *A Citizen's Guide to Radon: What It Is and What to Do About It*, OPA-86-004 (August).

Gerusky T. M. , 1986, "Pennsylvania's Radon Program," talk manuscript (unpublished) and supplementary table, March 1986.

Harley J. H., 1983, "Environmental Radioactivity-Natural," in: Environment Radioactivity, Proceedings of the Nineteenth Annual Meeting of the National Council on Radiation Protection and Measurements, Proceedings No. 5, p. 12, (Bethesda, MD:

NCRP).

ICRP (International Commission on Radiological Protection), 1977, "Recommendations of the International Commission on Radiological Protection," ICRP Publication 26, *Annals of the ICRP 1*, No. 3.

_____, 1977, "Problems Involved in Developing an Index of Harm," ICRP Publication 27, *Annals of the ICRP* 1, No. 4.

_____, 1981, "Limits for Inhalation of Radon Daughters by Workers," ICRP Publication 32, *Annals of the ICRP* 6, No. 1.

_____, 1985, "Statement from the 1985 Paris Meeting of the International Commission on Radiological Protection," *Radiation Protection Dosimetry* **11**, 134.

Jacobs D. G. , 1983, "Man-Made Sources Released to the Environment," in: Environmental Radioactivity, Proceedings of the Nineteenth Annual Meeting of the National Council on Radiation Protection and Measurements, Proceedings No. 5, p. 27, (Bethesda, MD: NCRP).

James A. C., 1987, "A reconsideration of cells at risk and other key factors in radon daughter dosimetry," in: *Radon and Its Decay Products*, ACS Symposium Series 331 (edited by Philip K. Hopke), pp. 400-418 (Washington, DC: American Chemical Society).

Kemeny J. G., Chairman, 1979, "Report of the President's Commission on the Accident at Three Mile Island," (New York: Pergamon), p. 34.

NAS (National Academy of Sciences/National Research Council), 1980, *The Effects on Populations of Exposure to Low Levels of Ionizing Radiation: 1980*, Report of the Committee on the Biological Effects of Ionizing Radiation (BEIR III) (Washington, DC: National Academy Press).

NCRP (National Council on Radiation Protection and Measurements), 1971, "Basic Radiation Protection Criteria," *NCRP Report No. 39* (Bethesda, MD: NCRP).

_____, 1984, "Exposures from the Uranium Series with Emphasis on Radon and Its Daughters," *NCRP Report No. 77* (Bethesda, MD: NCRP).

_____, 1984, "Evaluation of Occupational and Environmental Exposures to Radon and Radon Daughters in the United States," *NCRP Report No. 78* (Bethesda, MD: NCRP).

NEA (Nuclear Energy Agency, Organization for Economic Co-operation and Development), 1983, "Dosimetry Aspects of Exposure to Radon and Thoron Daughter Products," *Report by a Group of Experts* (Paris: OECD).

Nero A. V., Schwehr M. B., Nazaroff W. W. and Revzan K. L., 1986, "Distribution of airborne radon-222 concentrations in U.S. homes," *Science* 234, 992.

NN (Nuclear News), 1987, *Nuclear News Briefs*, Nuclear News 30, No. 1, p. 132.

USSR State Committee on the Utilization of Atomic Energy, 1986, *The Accident at the Chernobyl Nuclear Power Plant and Its Consequences*, Information Compiled for the IAEA Experts' Meeting, 25-29 August, Vienna.

GLOSSARY

Absolute risk: A risk to a population of an adverse health effect due to some injury (e.g., radiation dose), which is proportional to the magnitude of the injury to the population.

Aerosol: A system of solid or liquid particles which are (a) dispersed in a gaseous medium, (b) able to remain suspended in the gaseous medium for a long time relative to the time scale of interest, and (c) have a high surface area to volume ratio.

ALARA: An acronym standing for "as low as reasonably achievable"; expresses the principle that radiation exposure should be kept as far below regulatory limits as can be achieved with reasonable expense.

Alpha particle: The nucleus of the helium isotope of mass 4; emitted in the decay of some heavy radioactive nuclei.

Alveoli: The small sacs at the end of the air conduction system of the lung in which oxygen in the air is exchanged for carbon dioxide in the blood.

AMD: Activity mean diameter, the diameter of a sphere which contains the same uniform radioactivity per unit mass as the average radioactivity per unit mass in a collection of irregularly shaped particles.

Angstrom: One angstrom equals 10^{-10} meters. There are 10,000 angstroms in one micron (micrometer).

Arithmetic mean: The average value of a distribution of N items obtained by summing the individual magnitudes and dividing the total by N.

Attached fraction: The fraction of a short-lived radon daughter which is adsorbed onto or absorbed into the particles of the atmospheric aerosol. The attached fraction may be expressed either with respect to the actual activity of the radon daughter or with respect to the activity of the daughter at secular equilibrium.

Basal cells: The cells of the walls of the trachea and bronchial airways of the lung which divide to replenish the other cells of the airway walls.

Becquerel (Bq): The S.I. unit of rate of radioactive decay; 1 Bq equals 1 disintegration per second.

BEIR reports: A series of reports by the Committee on the Biological Effects of Ionizing Radiation of the National Academy of Sciences - National Research Council, dealing with the health effects of ionizing radiation.

Beta particle: An electron (with either negative or positive charge); emitted in the decay of some radioactive nuclei.

BPA: The Bonneville Power Administration, an agency of the Department of Energy which administers the distribution and sale of electric power generated mainly by the public hydroelectric facilities of the Pacific Northwest.

Bronchi: The subdividing airways which conduct air through the tracheo-bronchial region of the lung.

Bronchioles: The smaller bronchi.

Brownian motion: The random motion of small particles, for example those comprising the atmospheric aerosol, under the impact of the gas molecules of the atmosphere.

CFR: The Code of Federal Regulations, a compilation of the regulations set by federal agencies.

Cilia: Thread-like projections of specialized cells in the bronchial walls. The beating cilia propel foreign matter up to the throat to be swallowed and eliminated via the gastro-intestinal tract.

Cosmic rays: High energy radiations arriving at the earth from space. These radiations originate both from the sun and from beyond the solar system.

Crawl space: An area beneath some houses that separates the underflooring from the soil beneath.

Curie (Ci): a unit of rate of radioactive decay; 1 Ci equals 3.7×10^{10} disintegrations per second.

Daughter: A daughter or descendant atom is the atom produced by the decay of a radioactive parent atom.

Decay chain: A radioactive isotope and the series of radioactive daughters which are generated from it through a succession of radioactive decays. The chain ends when one of the daughters is nonradioactive.

Dose response curve: The curve of the relationship between the average dose of some toxic substance such as radiation, received by a large population, and the increase in the rate of some disease such as cancer.

Dose equivalent: See "rem".

Electron-volt (ev): Amount of kinetic energy acquired by an electron (or other particle carrying the same charge) when it is accelerated through a potential difference of one volt.

Emanation: The gas emitted from a solid or liquid. Radon was originally called "emanation" since it was emitted by radium.

Epithelium: The tissue of the walls of the trachea and bronchial airways composed of a basement membrane upon which are the basal cells, whose division replenishes the other cells of the epithelium, namely the goblet cells which secrete mucus and the ciliated cells whose cilia propel the mucous layer up to the throat.

Equilibrium factor (F): The ratio of the number of working levels actually present to the number of working levels which would be present if the short-lived daughters were in equilibrium with the radon present.

Equivalent equilibrium concentration (EEC): That concentration of radon in equilibrium with its short-lived daughters which would produce the same number of working levels as are actually present. The ratio of the EEC to the actual radon concentration is equal to the equilibrium factor, F.

Fluorspar: Calcium fluorite.

Gamma ray: Radiation, similar to x-rays and light, emitted in the decay of some radioactive nuclei.

Generation: Each airway segment in the lung divides to form (usually) two smaller airways. All of the airway segments which have arisen from the same number of divisions, starting from the trachea, is called a "generation". The trachea is commonly referred to as generation zero.

Gray: The S.I. unit of physical dose, i.e., the unit of deposition of energy in material due to the passage of ionizing radiation; 1 Gray = 1 joule per kilogram.

Ground water: The free water located within the soil and rocks that is not combined as water of hydration in minerals.

Half-life ($T_{1/2}$)*:* The time required for the number of radioactive nuclei present at any instant of time to be reduced by a factor of two, due to radioactive decay.

ICRP: The International Commission on Radiological Protection; an international group of experts who recommend limits for exposure to ionizing radiation to the international community.

Ionization chamber: A radiation detector which responds to the ionization current produced in the contained gas by the passage of ionizing radiation.

Ionizing radiation: Radiation with the ability to interact with and remove electrons from the atoms of material, leaving the atom ionized.

Jacobi-Eisfeld model: A dosimetric model for radon daughter dosimetry in the lung based on the Weibel "A" model.

James-Birchall model: A dosimetric model for radon daughter dosimetry in the lung based on consideration of both the Weibel and the Yeh-Schum lung models.

Laminar flow: Layered flow of a fluid (e.g., air) in which one layer slides smoothly over the other.

Latent period: A waiting period of time following some injury during which the ill effects of the injury are not yet apparent. In the case of radiogenic cancer, it is the period of time between the radiation exposure of a population and the first appearance of any resulting cancer in that population.

Linearity hypothesis: In the linearity hypothesis the assumption is made that the biological damage due to radiation is linearly proportional to the radiation dose.

Lognormal distribution: When the logarithms of some randomly distributed quantity have a normal (Gaussian) distribution, the distribution is referred to as lognormal. For a lognormal distribution, the geometric mean is the antilog of the arithmetic mean of the logarithmic values.

Median: The median of a distribution of values is that value at which half of them are smaller and half are larger.

Micro: A prefix indicating one one-millionth of a quantity (e.g., 1 microcurie = 10^{-6} curie).

Micron: One micron equals 10^{-6} meters, or 1 micrometer.

Mill tailings: The waste part of an ore after the valuable minerals have been removed at the separations mill. In a uranium mill, the tailings contain all of the radioactive daughters of uranium.

Milli: A prefix indicating one one-thousandth of a quantity (e.g., 1 millirem (mrem) = 10^{-3} rem.)

Nano: A prefix indicating one one-billionth of a quantity. (e.g., 1 nanocurie = 10^{-9} curie.)

Naso-pharygeal: Referring to the nose, throat and pharynx region of the respiratory system. Often abbreviated as N-P.

NCRP: The National Council on Radiation Protection and Measurements; a Congressionally chartered group of experts who are charged with studying the effects of exposure to ionizing radiation and recommending protective measures.

Noble gas: A noble gas is a gaseous element with negligible chemical reactivity. Helium, neon, argon, krypton, xenon and radon are the noble gases. Radon is the only radioactive noble gas.

Normal distribution: A term referring to the so-called "bell shaped curve" of randomly distributed quantities. This distribution is also referred to as a "Gaussian distribution".

NRC: Nuclear Regulatory Commission; also National Research Council.

Partial pressure: The contribution to the overall gas pressure by one of the gases in a mixture of gases.

Pico: A prefix indicating one one-trillionth of a quantity (e.g., 1 picocurie (pCi) = 10^{-12} Curie.)

Pitchblende: An ore of uranium oxide very rich in uranium.

Potential alpha energy: The amount of alpha-particle kinetic energy that can be dissipated within an atmosphere containing some particular mixture of short-lived daughters of radon, if they all decay there.

Precursor: As applied to radioactivity, a precursor of a radioactive isotope is a radioactive chain member which occurs earlier in the decay chain.

Pulmonary: As used here, refers to the smallest airways and alveolar sacs of the lung. The "deepest" part of the lung. Often abbreviated as P.

Quality factor (QF or Q): A modifying factor used to obtain a common basis for the human health hazard from different radiations. The value of the quality factor increases with increasing linear density of energy deposition in the tissue through which the radiation passes. See "Rem" and "Sievert".

Rad (rad): A traditional unit of physical radiation dose, i.e., the unit of deposition of energy in material due to the passage of ionizing radiation; 1 rad = 100 ergs per gram.

Radioactivity: The process by which an atom changes spontaneously into a different atom by the emission of an energetic particle from its nucleus.

Radium: A naturally occurring radioactive element whose decay produces radon. Radium is a member of the decay chain of uranium.

Radon daughters: The term "radon daughters" usually refers to the short-lived radioisotopes in the decay chain of radon down to lead-210. These are polonium-218 (RaA), lead-214 (RaB), Bismuth-214 (RaC), and Polonium-214 (RaC′).

Radon: A radioactive noble gas generated by the decay of radium.

Reading Prong: Descriptive name given to a geological feature which extends from the vicinity of Reading, Pennsylvania in a northeasterly direction through New Jersey into New York. Parts of it are characterized by very high levels of radon in buildings.

Regional dose factor: A multiplicative factor which converts the radiation dose equivalent received by some specific tissue into an effective whole body dose equivalent.

Relative risk: A risk of an adverse health effect due to some injury which is proportional both to the magnitude of the injury and to the usual rate of occurrence of the adverse health effect in the population at risk.

Rem (rem): A traditional unit of "dose equivalent" used to express on a common basis the health hazard from different kinds of radiations; the dose equivalent in rem is equal to the product of the quality factor (and other modifying factors, if used) and the physical dose in rad. (1 rem $= 10^{-2}$ sieverts.)

Secular Equilibrium: When a chain of radioactive daughters is produced by a long-lived radioactive parent, secular equilibrium is reached when the activity (in Ci or Bq) of each radioactive daughter is equal to the activity of the radioactive parent.

Sievert (Sv): The S.I. unit of "dose equivalent" is used to express on a common basis the health hazard from different kinds of radiations; the dose equivalent in sieverts is equal to the product of the quality factor and the physical dose in grays. (1 sievert = 100 rem.)

Somatic: Referring to the body cells of an organism. Somatic effects of radiation exposure appear in the irradiated organism itself, in contrast with genetic effects which appear in the descendants of the irradiated organism.

Specific activity: The amount of activity in a unit amount of material. Usually specified as amount of activity per unit mass, e.g., pCi/g.

Stochastic: Relating to a random process which has only a statistical chance of occurring. Radiation induced cancers have only a small chance of occurring and their incidence rate is a stochastic process.

Synergism: A term describing the combining of the effects of two kinds of injuries to produce a rate of adverse health effects larger than the sum of the effects of each when applied separately.

Trachea: The main airway of the respiratory system connecting the throat to the lungs.

Tracheo-bronchial: Referring to the larger airways of the lung from the trachea down to the bronchioli. Often abbreviated as T-B.

Turbulent flow: The flow of a fluid that overall is moving in a definite direction but any particular parcel of fluid follows an irregular chaotic path.

Unattached fraction (f_i): The ratio of the specific activity of RaX which is not attached to the atmospheric aerosol to the total specific activity of RaX which is present in any form in the atmosphere. ($i = 1, 2$, or 3, and X = A, B, or C for RaA, RaB, or RaC, respectively).

Unattached fraction (f_p): The ratio of the number of working levels contributed by the unattached RaA, RaB, and RaC to the total number of working levels.

Unattached fraction (f_x): the ratio of the specific activity of RaX which is not attached to the atmospheric aerosol to the total specific activity of RaX which would be present in equilibrium with the specific activity of radon present. (X = A, B, or C).

Weatherization: The process of insulating a house and/or reducing pathways for outside air to leak into a house so as to reduce the heating and cooling requirements.

Weibel "A" model: A geometrical model of the lung in which every airway divides symmetrically into two airways of the next generation.

Whole-body dose: The dose received from a radiation exposure which irradiates the entire body.

Working level month (WLM): A unit of exposure to short-lived radon daughters equal to the exposure to one WL for one working month of 170 hours.

Working level (WL): One working level is that amount of potential alpha-particle energy dissipated in air by the short lived daughters in equilibrium with 100 pCi/ℓ of radon. One WL is equal to 130,000 MeV of alpha-particle energy deposited per liter of air.

X-ray: Radiation similar to gamma-rays and light, emitted during rearrangements of the inner electrons surrounding an atom.

Yeh-Schum model: A geometrical model of the lung in which each of the five lobes of the lung has its own pattern of regular, but not necessarily symmetric, airway divisions from generation to generation.

SUBJECT INDEX